Improve Your Maths!

Essential Maths for Students

Series Editors: Anthony Croft, Department of Mathematical Sciences, Loughborough University, and Robert Davison, Department of Mathematical Sciences, De Montfort University, Leicester

Also available
Foundation Maths, 2nd edition
Croft and Davidson

Engineering Maths
Mustoe

Maths for Computing and Information Technology
Giannasi and Low

Maths and Statistics for Business
Lawson, Hubbard and Pugh

Engineering Systems: Modelling and Control
Hargreaves

Essential Maths for Students

Improve Your Maths!

A refresher course

**Gordon Bancroft
and Mike Fletcher**

 **Addison-Wesley**

Harlow, England • Reading, Massachusetts • Menlo Park, California
New York • Don Mills, Ontario • Amsterdam • Bonn • Sydney
Singapore • Tokyo • Madrid • San Juan • Milan • Mexico City
Seoul • Taipei

Addison Wesley Longman Limited
Edinburgh Gate
Harlow
Essex
CM20 2JE
England

and Associated Companies throughout the world.

Cover figure Pantek Arts, Maidstone, Kent

Typeset by 32
Typeset in 10/12 Times
Produced by Longman Singapore Publishers (Pte) Ltd.
Printed and bound by CPI Antony Rowe, Eastbourne

First printed 1998
Transferred to digital print on demand, 2006

ISBN 0-201-331306

British Library Cataloguing-in-Publication Data
A catalogue record for this book is available from the British Library

To Anne and Sheila

Contents

Preface

This book aims to help all students who intend to start or who have already started a course in business studies, the social sciences or other subject areas that require a student to be reasonably numerate. It will improve, refresh and add to students' basic numeracy and mathematical ability. It can be used both as an introductory text before a course begins and throughout a course wherever a particular mathematical skill or concept causes difficulty. Individuals who are not students, but who wish to develop numeracy skills vital to everyday life, may also find this text a valuable aid.

Concentrating on basic numeracy, statistical, graphical and simple algebraic skills, it can be used without supervision. Clear and concise, with a variety of illuminating examples, it is extremely practical. The exercises, with solutions provided, make this book ideal for self-study purposes. Indeed, anyone just wishing to improve their maths skills will find this book a great help.

Throughout, the book introduces the necessary techniques through the medium of worked examples. Each chapter begins with a statement of the objectives of that chapter, describing the skills and abilities that the reader should acquire after successfully reading the chapter. Most sections conclude with self-assessment questions, which provide a quick guide to how well the material has been understood. The self-assessment questions are followed by a set of exercises on the material of each section, to which complete answers are provided at the end of the book. Each chapter is followed by a set of test exercises from which a tutor can set assignments or tests.

The book makes continued reference to the use of the calculator, an important tool in today's society. In many of the numeracy chapters the reader is shown, step by step, how to use the calculator in obtaining answers to problems. However, the text does not rely on the reader having access to sophisticated computer packages that are available in today's computer-literate world.

It is important to realize that to understand a piece of mathematics fully requires considerable application by attempting and successfully completing a number of examples and exercises. As the reader does more and more practice the work should become easier and thinking numerically should become more natural.

Gordon Bancroft and Mike Fletcher

Acknowledgements

We have been unable to trace the copyright holders of sources for Table 5.7 and Table 5.11, and would be grateful for any information that will enable us to do so.

Section A
Numbers

1 Arithmetic operations

Objectives

After reading this chapter, you should be able:

- to handle numerical problems dealing with addition, subtraction, multiplication and division of whole numbers
- to perform calculations involving negative whole numbers
- to make use of brackets in numerical situations
- to round numbers to required levels of accuracy

1.1 Introduction

In all walks of life, whether at work, at home or socially, an ability to handle numbers and to perform basic arithmetic is an essential requirement in today's society. Indeed, it is as important to understand basic arithmetic as it is to read and write. In this opening chapter the basic operations of arithmetic, namely addition, subtraction, multiplication and division, are described and applied to problems involving whole numbers.

1.2 Numerical problems involving whole numbers

Although you may feel you have confidence in handling whole numbers and you feel that they will cause little difficulty, do you know how to compute, for example:

$$19 - 4 \times 2?$$

The correct answer is 11 but did you get the answer 30? Many of you will. If, perhaps, brackets had been used it would have made the order of operations more clear. In this case

$$19 - (4 \times 2) = 19 - 8 = 11$$

The answer 30 is, of course, the correct answer to the problem

$$(19 - 4) \times 2$$

which has the same numbers and the same operation signs but the order of operations is different. Therefore it is clearly sensible to use brackets to emphasize the order of the arithmetic operations and hence to remove any ambiguity that may exist. An essential rule in mathematics, not just arithmetic, is that expressions within brackets are evaluated first. So

$$(12 \times 7) - (8 \times 5) = 84 - 40 = 44$$

However, this does not help with the initial problem, $19 - 4 \times 2$, which contains no brackets. How would your calculator cope with this problem?

Take another example

$$8 + 4 \times 3$$

and key into your calculator

$$\boxed{8}\ \boxed{+}\ \boxed{4}\ \boxed{\times}\ \boxed{3}\ \boxed{=}$$

Most good calculators will give 20 for the answer as they perform the multiplication operation before the addition. This is standard practice in mathematics and you should always follow this pattern, so

$$8 + 4 \times 3 = 8 + 12 = 20$$

Unfortunately some calculators do work out the expression from left to right giving the answer of 36 in this example. Therefore it is a sensible idea to become familiar with the way that your calculator performs this process. From now on, perform the multiplication (or division) operation before addition (or subtraction). To make this more clear the convention of using brackets to replace the multiplication sign is often used. For example, $12 \times 4 - 5 \times 8$ may be rewritten in the form $12(4) - 5(8)$ to give 8 as the answer. This convention of using brackets instead of the multiplication sign will be used throughout this text. Besides making the order of operation more clear it also avoids the confusion between the multiplication sign $\times$ and the letter x.

If a problem contains only pluses or minuses (or only multiplication and division) there is not usually any confusion as the expression is evaluated from left to right. For example:

$$12 + 3 - 7 + 2 = 15 - 7 + 2 = 8 + 2 = 10$$

Use your calculator to confirm this answer.

KEY POINT

Consequently the following summary can be used to evaluate a stated arithmetic problem:

1. Use brackets to clarify the correct order of operation.

2. Evaluate expressions within brackets first.

3. Perform multiplication (and division) before addition (and subtraction).

4. Work from left to right if the expression contains only addition and subtraction (or multiplication and division only).

These rules are illustrated now in an example that describes a situation in everyday life.

Worked example

1.1   Chris and David go out one evening with their wives, Mary and Julie, to a local public house, The Three Crowns. The price list for drinks is displayed by the bar and part of it can be seen in Table 1.1.

Table 1.1.
Drinks price
list

Drink	Price (pence)
Beer–pint	150
Beer–half pint	77
Lager–pint	160
Lager–half pint	82
Gin	140
Whisky	150
Rum	170
Lime additive	10
Tonic	50
Ginger ale	60
Orange juice	88

Chris ordered the first round of drinks, two pints of lager and lime, one gin and tonic, and one orange juice. Find the cost of this round of drinks.

Solution   The cost of the first round can be expressed by

$$2(160 + 10) + (140 + 50) + 88 \text{ pence}$$

You can see that the cost of two lager and limes is given by $2(160 + 10)$, showing the usefulness of the brackets notation. Note that the expression $2 \times 160 + 10 = 330$ gives a different answer to the correct expression $2(160 + 10)$. The total cost for this round of drinks is, again

$$2(160 + 10) + (140 + 50) + 88$$

Evaluating the terms inside the brackets gives

$$2(170) + 190 + 88$$

Performing the multiplication before any addition

$$340 + 190 + 88$$

Now working from left to right

$$530 + 88 = 618 \text{ pence (or £6.18)}$$

David bought the second round of drinks; two pints of beer (the lager was not to their taste), one gin and tonic, and a ginger ale. David paid for his round with a £10 note, so the amount of change he receives is given by the following arithmetic expression

$$1000 - [2(150) + (140 + 50) + 60] \text{ pence}$$

Using the rules describing the correct order of operation, David's actual change is

$$1000 - [2(150) + 190 + 60] = 1000 - [300 + 190 + 60]$$
$$= 1000 - [550]$$
$$= 450 \text{ pence (or £4.50)}$$

Worked example

1.2   A quarterly telephone bill includes both a standing charge of £19.50 plus a further charge of 8.5 pence per metered unit. Give an arithmetic expression for the total bill if 250 meter units are used. Evaluate the expression.

Solution   Total bill (in pence) $= 1950 + 250(8.5)$
$$= 1950 + 2125$$
$$= 4075 \text{ pence}$$

The telephone bill will be £40.75.

Self-assessment question 1.2

1. Without using a calculator work out
   (a) $8 \times 3 - 4 \times 2$
   (b) $18 \div 3 + 2 - 5$
   (c) $15 \times 2 - 1 \times 9$
   (d) $(3 \times 4) - (8 \times 1)$
   (e) $\{[(3 \times 6) - 4] \times 3\}$
   (f) $(6 \div 3) \times (2 + 1)$

Exercise 1.2

1. You meet four friends in a coffee bar and offer to buy them all coffee (you have one as well). Coffee costs 42p per cup. What will be the total cost of the coffees?

2. Work out the following:
   (a) $975 + 329$
   (b) $557 - 294$
   (c) $108 \div 12$
   (d) 6×31

3. In the first two hours of a shift an operator makes 16 units of output per hour. In the next three hours the operator makes 14 units per hour. In the final three hours 11 units are made each hour. How many units are made altogether in the eight hour shift?

4. Place brackets in the following expressions to make them correct:
   (a) $4 \times 12 - 5 + 2 = 30$
   (b) $3 \times 11 + 6 - 2 = 45$
   (c) $9 \times 10 - 4 + 6 = 92$

5. A typist finds that she can get 13 words in one line and 35 lines of typing on one page. If she types a manuscript consisting of 21 840 words, how many pages will she type?

1.3   Negative numbers

Arithmetic operations are further complicated if some of the numbers are negative. A daily weather report might give a maximum daytime temperature of 14 °C and might also give a minimum temperature on that day of −4 °C. The difference between two numbers is computed by subtracting the smaller number from the larger one, so the difference between these two quoted temperatures is given by

$$14 - (-4)$$

The answer is 18 °C, which can be seen more clearly by observing Figure 1.1.

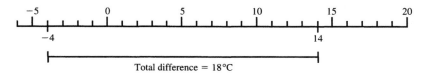

Figure 1.1.
Temperature differences

The operation of a bank account may be a further situation which requires the application of both positive and negative numbers. Here the value of a credit can be treated as a positive number, whereas the amount of a cheque or withdrawal is the negative number.

Worked example

1.3   On 1 June Mrs Worth had a balance of £255 in her account.

(a) During June she paid in amounts of £73 and £82 and wrote cheques to the value of £67, £96, £87, £92 and £82. What is the balance of her account at the end of June?

(b) During July of the same year there was one credit to the value of £120, but she wrote three cheques, £50 each. Determine the balance of the account at the end of July.

(c) How much must Mrs Worth pay into her account at the end of July to pay off her overdraft?

Solution   (a) As the values of the cheques are to be considered to be negative numbers, the balance at the end of June is given by the expression

$$255 + (73) + (82) + (-67) + (-96) + (-87) + (-92) + (-82)$$

This can be evaluated directly on a calculator using the key sequence

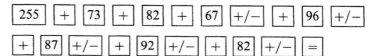

where the $+/-$ button changes the sign of the number displayed and is therefore useful when inputting negative numbers. The answer displayed is -14, indicating that Mrs Worth is £14 overdrawn at the end of June. It is useful to note that the addition of a negative number is equivalent to subtracting a positive number. Therefore her bank balance could have been calculated using the arithmetic expression

$$255 + (73 + 82) - (67 + 96 + 87 + 92 + 82) = 255 + 155 - 424 = -14$$

(b) As the bank balance at the beginning of July was $-£14$, the balance at the end of the month is

$$(-14) + (120) + 3(-50) = -14 + 120 - 3(50)$$
$$= -14 + 120 - 150$$
$$= -44$$

Now the account is overdrawn by £44.

(c) In order to pay off her overdraft she must add

$$0 - (-44) = 44$$

that is, £44 to her account.

In this part note that subtracting a negative number is the same as adding a positive number:

$$0 - (-44) = 0 + 44 = 44$$

KEY POINT

The above example illustrates the following rules when adding or subtracting negative numbers:

1. Addition of a negative number is the same as subtraction of the equivalent positive number.
2. Subtraction of a negative number is the same as addition of the equivalent positive number.

What happens when we multiply using negative numbers? Use your calculator to evaluate the expressions in the next example taking care to write down the sign (positive or negative) of your answer. Remember to input the negative number by using the $\boxed{+/-}$ key on your calculator.

Worked example

1.4   (a) $7(-3)$ (b) $(-4)(-5)$ (c) $(-12)(4)$

Solution   (a) The key sequence $\boxed{7}$ $\boxed{\times}$ $\boxed{3}$ $\boxed{+/-}$ $\boxed{=}$ yields an answer of -21.

(b) The key sequence $\boxed{4}$ $\boxed{+/-}$ $\boxed{\times}$ $\boxed{5}$ $\boxed{+/-}$ $\boxed{=}$ gives an answer of $+20$.

(c) The key sequence $\boxed{12}$ $\boxed{+/-}$ $\boxed{\times}$ $\boxed{4}$ $\boxed{=}$ gives -48.

KEY POINT

The following rules apply to multiplication using negative numbers:

1. (positive number) (negative number) = negative number; or $(+)(-) = (-)$.
2. (negative number) (negative number) = positive number; or $(-)(-) = (+)$.
3. (negative number) (positive number) = negative number; or $(-)(+) = (-)$.

In the same way the following rules apply involving division of and division by negative numbers:

1. (positive number) $\div$ (negative number) = negative number; or $(+) \div (-) = (-)$.
2. (negative number) $\div$ (negative number) = positive number; or $(-) \div (-) = (+)$.
3. (negative number) $\div$ (positive number) = negative number; or $(-) \div (+) = (-)$.

These rules always apply for both multiplication and division. Perhaps the second of these rules is difficult to comprehend, but in speech a double negative statement, such as 'not unusual', may imply a positive statement, such as 'usual'. Now answer the next example without using your calculator.

Worked example

1.5   Evaluate

(a) $8(-4)$ (b) $(-5)(-2)$ (c) $(-8) \div (-4)$ (d) $(56) \div (-7)$

Solution   (a) As $8(4) = 32$ and multiplication here is between a positive number and a negative number, we have

$$8(-4) = -8(4) = -32$$

(b) Multiplication of two negative numbers gives a positive answer

$$(-5)(-2) = 5(2) = 10$$

(c) Division of two negative numbers also gives a positive answer

$$(-8) \div (-4) = 8 \div 4 = 2$$

(d) Here $(56) \div (-7) = -(56 \div 7) = -8$

Self-assessment question 1.3

1. If one negative number is divided by another negative number, is the result negative or positive?

Exercise 1.3

1. Work out
   (a) $(52 - 7) - (-2 + 1)$
   (b) $(90 + 14) \div (6 - 4)$
   (c) $8(7 - 2) \div 2(5 - 3)$

2. You have £36 left in your bank account. However, the bank has just told you that you owe £40 in bank charges and has deducted this from your account. How much is left in your account?

3. Work out
   (a) $10 + (-6)$
   (b) $-10 + (-6)$
   (c) $-10 - (-6)$
   (d) $-10 - (+6)$
   (e) $10 - (-6)$

4. Mr Smith's current account was £53 overdrawn. After an amount was paid into this account the balance was £75 in credit. How much was paid into this account?

5. Work out the following:
   (a) $15 \times (-3)$
   (b) $(-15) \times (-3)$
   (c) $(-15) \div (-3)$
   (d) $15 \div (-3)$

6. A rock climber started climbing in the morning at 120 metres below a camping position and ended up at the end of the day 60 metres above the camping position. What distance did the rock climber achieve during the day?

1.4	The rounding of numbers

A further useful skill when dealing with numbers is that of rounding a number to the degree of accuracy appropriate to the situation in hand. Often this has the additional advantage of making the general size of the number more comprehensible. For example, you may have a large win on the Lottery which amounts to £253 427.93p. (If so, why are you reading this book?) It is quite likely that this figure will be reported in the national press as £250 000. From the reader's point of view this latter figure gives a good idea of the size of the win; it provides a good approximation to the actual win and is easy to remember. This type of approximation is often carried out with large numbers, or with numbers that have a lot of digits, 372.642 934 say. In the Lottery example the win has been rounded to the nearest ten thousand, or correct to two significant figures (as there are two digits to the left of the zeros).

There are many situations where it is advantageous to estimate the value of a quantity. For example, you may wish to know approximately how long it would take to journey from Stoke-on-Trent to London. Using the AA-recommended route the distance by road is 156 miles but it is unlikely that you are travelling from the centre of Stoke-on-Trent to the centre of London so it may be sensible to round the distance to the nearest ten miles. As 156 miles is nearer to 160 miles than 150 miles, the distance between the two cities is 160 miles to the nearest 10 miles.

Worked example

1.6   Table 1.2 shows the distance, in miles, between major cities within Great Britain. The distances are given to the nearest mile and are measured along the normal AA-recommended route.

(a) Determine the distance between the following pairs of cities rounded to the nearest 10 miles:
  (i)   Exeter and Cardiff.
  (ii) Edinburgh and Liverpool.
  (iii) London and Cardiff.
(b) What is the distance between London and Edinburgh to the nearest 100 miles?
(c) What is the distance between London and Liverpool rounded to one significant figure?

Table 1.2.
Mileage chart for certain cities

London				
405	Edinburgh			
155	337	Cardiff		
210	222	200	Liverpool	
170	446	119	250	Exeter

Solution   (a) (i) The distance between Exeter and Cardiff is 119 miles which is nearer to 120 miles than 110 miles, so the answer is 120 miles.

(ii) 222 miles is closer to 220 miles than 230 miles, so the answer is 220 miles.

(iii) The distance between London and Cardiff is 155 miles, which is exactly halfway between 150 miles and 160 miles. Convention states that if this happens then the number should be rounded up rather than down. This gives an answer of 160 miles.

(b) 405 miles is nearer to 400 miles than 500 miles, so choose 400 miles.

(c) The numbers 200 and 300 both have one significant figure (one number to the left of the zeros) and the distance 210 miles is nearer to 200 than 300. The answer is 200.

Self-assessment question 1.4

1. Explain why it is useful to round numbers. What is lost when numbers are rounded?

Exercise 1.4

1. 5695 people live in Ayford and 9406 people live in Beeton. Round these figures to the nearest thousand.

2. Write the following numbers correct to three significant figures:
   (a) 5563
   (b) 147 296
   (c) 1133

3. Round these numbers to the nearest 100:
   (a) 237
   (b) 534 871
   (c) 560
   (d) 42

4. You read in the paper that a particular company made a profit last year of £3 873 584. What would this be to the nearest
   (a) hundred pounds?
   (b) ten thousand pounds?

5. Jean's car does 43 miles to the gallon. Each week she buys 11 gallons. How far does she travel each week, to the nearest 10 miles?

Test and assignment exercises 1

1. Evaluate the following expressions:

   (a) $7 + 5(2)$
   (b) $3 + 7(4)$
   (c) $5(3) + 6$

2. A rock climber started climbing 50 metres below a ledge and ended up 35 metres above the ledge. Express the distance climbed arithmetically and evaluate it.

3. Without using a calculator evaluate
   (a) $((-3) + 7) \div 2$
   (b) $((-2) - (-4))3$
   (c) $((-5) + (-3)) \div (-4)$

4. The attendance at a football match was
   44 821.
   (a) Round the attendance to the nearest
       hundred.
   (b) Round the attendance to the nearest
       thousand.

5. Use your calculator to compute
   (a) $350 + 5(25)$
   (b) $80(4) + 10$
   (c) $24(5) + 18(7)$

6. During the first quarter of 1995 there
   were 50 963 marriages in England and
   Wales. What is this value to the nearest
   thousand?

7. Use your calculator to compute
   (a) $(-40) + (-5)16$
   (b) $((-25) + (-365)) \div (-13)$

8. The deepest place in the Pacific Ocean is
   36 000 feet below sea-level in the
   Marianas Trench near Guam. The
   highest mountain in the Himalayas is
   Mount Everest at 29 000 feet above sea-
   level. What is the difference in their
   'heights'.

9. Mrs Prince's current account was £52
   overdrawn. After an amount was paid
   into this account the balance was £36 in
   credit. How much was paid into the
   account?

10. Evaluate $83(36) \div (6 + 4(3))$

11. Your monthly gross salary is £645 and
    you pay £124 in taxes and £43 in
    National Insurance each month.
    Calculate your take-home pay.

12. A family bought a dining room table,
    priced at £323, four dining chairs,
    priced at £62 each, and two carver
    chairs, £83 each.
    (a) Write a mathematical expression
        that determines the total value of
        dining room furniture.
    (b) Determine the total expenditure.
    (c) Round your answer to (b) to the
        nearest £10.

2 Fractions and decimals

Objectives

After reading this chapter, you should be able:

- to express a fraction in its simplest form
- to add, subtract, multiply and divide fractions
- to use a calculator to carry out calculations involving decimal numbers
- to demonstrate skills in rounding decimal numbers

2.1 Introduction

In Chapter 1 attention was restricted to arithmetic operations involving whole numbers. However, a glance at any newspaper may reveal, say, that yesterday's rainfall was 0.65 mm, or that the price of a particular article of clothing is 'one quarter off', or that the exchange rate is 1.615 dollars to the pound. All of these pieces of information contain numbers that are part of a whole. It is important to understand this concept of fractions, whether it be common fractions or decimal 'fractions'.

Are you ill, or just a fraction?

2.2   Fractions

A *common fraction* is a number written like

$$\frac{3}{5}$$

where the number above the line is called the *numerator* and the number below the line is called the *denominator*. The fraction $\frac{3}{5}$ can be envisaged by splitting the whole into five equal parts and taking three of these parts; see Figure 2.1.

Figure 2.1.
Representation of
three-fifths

$\frac{3}{5}$ Three-fifths

By considering similar diagrams (see Figure 2.2) it can be seen that the fractions $\frac{2}{4}$ and $\frac{1}{2}$ are the same. We conclude that

$$\frac{2}{4} = \frac{1}{2}$$

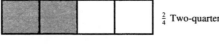
$\frac{2}{4}$ Two-quarters

Figure 2.2
Representation of
one-half

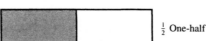
$\frac{1}{2}$ One-half

In fact there are many different ways of writing one-half:

$$\frac{1}{2} = \frac{2}{4} = \frac{3}{6} = \frac{4}{8} = \frac{5}{10} = \frac{8}{16}$$

These are all equal or equivalent fractions and can be formed by multiplying both the numerator and denominator of the fraction by the same constant. It is usually best to reduce a fraction to its simplest form. To do this, divide both the numerator and denominator by the same whole number. When no number other than one will divide into both the numbers, the fraction is said to be in its *lowest form*.

KEY POINT

A fraction is in its lowest form when no whole number other than one will divide into both the numerator and the denominator of the fraction.

Worked example

2.1   Convert the following fractions to their lowest terms:

(a) $\frac{3}{12}$ (b) $\frac{16}{28}$ (c) $\frac{5}{8}$

Solution   (a) $\frac{3}{12} = \frac{1}{4}$ as both the numerator and denominator of the original fraction can be divided by 3.

(b) $\frac{16}{28} = \frac{4}{7}$. Both numbers can be divided by 4.

(c) $\frac{5}{8}$ is in its lowest form. Although the denominator can be divided by 2 and 4 the numerator cannot. Hence it is in its lowest form.

In order to add or subtract two or more fractions they must have the same denominator. Therefore the first step when carrying out this operation is to find the common denominator of all the fractions.

Worked example

2.2   Evaluate

(a) $\frac{1}{5} + \frac{3}{5}$ (b) $\frac{1}{2} + \frac{1}{3} - \frac{1}{6}$ (c) $\frac{3}{4} + \frac{5}{6}$

Solution   (a) $\frac{1}{5}$ and $\frac{3}{5}$ both have a denominator equal to 5. Therefore these fractions can be added immediately:

$$\frac{1}{5} + \frac{3}{5} = \frac{1+3}{5} = \frac{4}{5}$$

(b) The smallest number divisible by 2, 3 and 6 is 6 so the fractions are converted to equivalent fractions with denominator equal to 6:

$$\frac{1}{2} + \frac{1}{3} - \frac{1}{6} = \frac{3}{6} + \frac{2}{6} - \frac{1}{6} = \frac{3+2-1}{6} = \frac{4}{6}$$

In its lowest form the answer is $\frac{2}{3}$.

(c) $\frac{3}{4}$ and $\frac{5}{6}$ have denominators equal to 4 and 6 respectively. The smallest number divisible by both 4 and 6 is 12. The two fractions need to be rewritten so that they are equivalent fractions with denominator equal to 12.

$$\frac{3}{4} = \frac{9}{12} \text{ and } \frac{5}{6} = \frac{10}{12} \text{ so}$$

$$\frac{3}{4} + \frac{5}{6} = \frac{9}{12} + \frac{10}{12} = \frac{19}{12}$$

The answer $\frac{19}{12}$ is an *improper fraction* (as the numerator is greater than the denominator) and can be written as the mixed number, $1\frac{7}{12}$. A *mixed number* is a combination of a whole number and a fraction.

KEY POINT

To add and subtract fractions we first express them all with the same denominator.

Worked example

2.3 A cake is split into 12 equal portions. Four portions are removed and eaten. What fraction of the cake is left?

Solution The fraction of the cake remaining is

$$1 - \frac{4}{12} = \frac{12}{12} - \frac{4}{12} = \frac{12-4}{12} = \frac{8}{12}$$

In its lowest form this is $\frac{2}{3}$.

Worked example

2.4 Sixty people were interviewed about their current political opinions. Each of the respondents was asked to name the political party they would vote for if there was a general election on the following day. The results are shown in Table 2.1.

(a) What fraction of the respondents would vote for the two traditional British political parties, Conservative and Labour?
(b) What fraction of the respondents would vote for the three leading political parties?
(c) Which of the parties has the greatest support in this survey?
(d) Determine the fraction of those questioned who intend to vote.
(e) Check that the fractions given in the question add to 1.

Table 2.1.
Survey on political
opinion

Political party selected	Fraction of respondents
Labour	$\frac{1}{3}$
Liberal Democrats	$\frac{3}{10}$
Conservative	$\frac{4}{15}$
Ecology	$\frac{1}{20}$
Other parties	$\frac{1}{30}$
Will not vote	$\frac{1}{60}$

Solution (a) $\frac{1}{3} + \frac{4}{15} = \frac{5}{15} + \frac{4}{15} = \frac{5+4}{15} = \frac{9}{15} = \frac{3}{5}$

(b) $\frac{1}{3} + \frac{3}{10} + \frac{4}{15} = \frac{10}{30} + \frac{9}{30} + \frac{8}{30} = \frac{10+9+8}{30} = \frac{27}{30} = \frac{9}{10}$

(c) Fractions can only really be compared if their denominators are the same.

Labour	$\frac{1}{3} = \frac{20}{60}$
Liberal Democrats	$\frac{3}{10} = \frac{18}{60}$
Conservative	$\frac{4}{15} = \frac{16}{60}$
Ecology	$\frac{1}{20} = \frac{3}{60}$

The largest of the numerators is 20, so the Labour party has the greatest support in this survey.

(d) $1 - \frac{1}{60} = \frac{60}{60} - \frac{1}{60} = \frac{60-1}{60} = \frac{59}{60}$

(e) The six fractions added together

$$\frac{1}{3} + \frac{3}{10} + \frac{4}{15} + \frac{1}{20} + \frac{1}{30} + \frac{1}{60} = \frac{20}{60} + \frac{18}{60} + \frac{16}{60} + \frac{3}{60} + \frac{2}{60} + \frac{1}{60} = \frac{60}{60} = 1$$

Worked example

2.5   A small company consists of 40 employees. Of the employees, 15 are female, 16 are aged under 30, and 32 work in the Production Department. Express as fractions:

(a) the proportion of production workers;
(b) the proportion of male workers;
(c) the proportion aged under 30.

Solution   (a) 32 out of the 40 employees are in the Production Department. Consequently the required fraction is

$\frac{32}{40}$

which in its simplest form is $\frac{4}{5}$.

(b) As 15 of the employees are female, the remainder $40 - 15 = 25$ are male. The proportion of male workers is

$\frac{25}{40} = \frac{5}{8}$

(c) The proportion aged under 30 is

$\frac{16}{40}$ or in its lowest form, $\frac{2}{5}$

Many good calculators are capable of performing calculations involving fractions directly. This involves using the $\boxed{a\frac{b}{c}}$ button. For example

$\frac{2}{3} - \frac{1}{6}$

can be worked out by using the key sequence:

$\boxed{2}$ $\boxed{a\frac{b}{c}}$ $\boxed{3}$ $\boxed{-}$ $\boxed{1}$ $\boxed{a\frac{b}{c}}$ $\boxed{6}$ $\boxed{=}$

The answer is one-half, given as 1⌋2, on most calculators.

In a similar way,

$1\frac{3}{4} + 2\frac{5}{8}$

is worked out by using the key sequence:

$\boxed{1}$ $\boxed{a\frac{b}{c}}$ $\boxed{3}$ $\boxed{a\frac{b}{c}}$ $\boxed{4}$ $\boxed{+}$ $\boxed{2}$ $\boxed{a\frac{b}{c}}$ $\boxed{5}$ $\boxed{a\frac{b}{c}}$ $\boxed{8}$ $\boxed{=}$

giving an answer of $4\frac{3}{8}$ (given as 4⌋3⌋8 on the calculator).

Worked example

2.6 Work out the following expressions involving fractions using your calculator:

(a) $\frac{3}{4} - \frac{1}{3}$ (b) $\frac{4}{7} + \frac{1}{5} - \frac{1}{3}$ (c) $1\frac{1}{3} + 1\frac{1}{2}$

Solution (a) The key sequence is $\boxed{3}$ $\boxed{a\frac{b}{c}}$ $\boxed{4}$ $\boxed{-}$ $\boxed{1}$ $\boxed{a\frac{b}{c}}$ $\boxed{3}$ $\boxed{=}$, which gives $\frac{5}{12}$ as the answer. By hand the answer is found in the following way:

$$\frac{3}{4} - \frac{1}{3} = \frac{9}{12} - \frac{4}{12} = \frac{5}{12}$$

(b) The key sequence is $\boxed{4}$ $\boxed{a\frac{b}{c}}$ $\boxed{7}$ $\boxed{+}$ $\boxed{1}$ $\boxed{a\frac{b}{c}}$ $\boxed{5}$ $\boxed{-}$ $\boxed{1}$ $\boxed{a\frac{b}{c}}$ $\boxed{3}$ $\boxed{=}$, or, by hand

$$\frac{4}{7} + \frac{1}{5} - \frac{1}{3} = \frac{60}{105} + \frac{21}{105} - \frac{35}{105} = \frac{46}{105}$$

(c) The key sequence is $\boxed{1}$ $\boxed{a\frac{b}{c}}$ $\boxed{1}$ $\boxed{a\frac{b}{c}}$ $\boxed{3}$ $\boxed{+}$ $\boxed{1}$ $\boxed{a\frac{b}{c}}$ $\boxed{1}$ $\boxed{a\frac{b}{c}}$ $\boxed{2}$ $\boxed{=}$, giving $2\frac{5}{6}$ as the answer:

$$1\frac{1}{3} + 1\frac{1}{2} = \frac{4}{3} + \frac{3}{2} = \frac{8}{6} + \frac{9}{6} = \frac{17}{6} = 2\frac{5}{6}$$

To multiply fractions there is *no* need to rewrite the fractions so that they have the same denominator. All you need to do is multiply the numerators together and multiply the denominators together. The resulting fraction is then reduced to its lowest form. For example,

$$\frac{8}{9}\left(\frac{3}{16}\right) = \frac{8(3)}{9(16)} = \frac{24}{144} = \frac{1}{6}$$

KEY POINT | Multiplication of two fractions can be carried out by multiplying their numerators to form a new numerator, and then multiplying their denominators to form a new denominator.

In order to simplify this operation you can divide across the fractions using a common number, a process called *cancellation*. In the above example both 8 and 16 can be divided by 8 and both 3 and 9 can be divided by 3, giving

$$\frac{8}{9}\left(\frac{3}{16}\right) = \frac{\cancel{8}^{1}(\cancel{3})^{1}}{\cancel{9}_{3}(\cancel{16})_{2}} = \frac{1(1)}{3(2)} = \frac{1}{6}$$

Even if one of the numbers is a whole number the procedure remains the same, since the whole number can be expressed as a fraction by giving it a denominator of 1. For example,

$$\frac{2}{3}(12) = \frac{2}{3}\left(\frac{12}{1}\right) = \frac{2(\cancel{12})^{4}}{\cancel{3}_{1}(1)} = \frac{8}{1} = 8$$

The word 'of' has the same meaning as 'multiplied by' in arithmetic. So the above calculation is equivalent to finding two-thirds of 12.

To divide one fraction by another, the second fraction (the *divisor*) is turned 'upside down' (the numerator becomes the denominator and the denominator becomes the numerator) and the two fractions are then multiplied together. Although this sounds a little complicated, it is not! For example

$$\frac{4}{15} \div \frac{2}{5} = \frac{4}{15}\left(\frac{5}{2}\right) = \frac{\cancel{4}^2(\cancel{5})^1}{\cancel{15}_3(\cancel{2})_1} = \frac{2(1)}{3(1)} = \frac{2}{3}$$

In this example $\frac{2}{5}$ is turned 'upside down' to give $\frac{5}{2}$ which is then multiplied by $\frac{4}{15}$, giving $\frac{2}{3}$ as the answer. In the same way

$$6 \div \frac{3}{5} = \frac{6}{1}\left(\frac{5}{3}\right) = \frac{\cancel{6}^2(5)}{1(\cancel{3})_1} = \frac{2(5)}{1(1)} = \frac{10}{1} = 10$$

KEY POINT

When dividing by a fraction, the fraction is turned 'upside down' and then multiplied.

Worked example

2.7   Evaluate the following expressions involving multiplication or division of fractions:

(a) $\frac{7}{10} \times \frac{5}{6}$ (b) $\frac{8}{15} \div \frac{2}{5}$ (c) $\frac{3}{5} \times \frac{2}{3} \div \frac{4}{5}$

(d) $1\frac{3}{5} \times 1\frac{1}{2}$ (e) $3\frac{1}{8} \div 1\frac{1}{4}$

Solution   (a)   The calculation can be carried out directly on most calculators using the key sequence:

$$\boxed{7}\ \boxed{a\frac{b}{c}}\ \boxed{10}\ \boxed{\times}\ \boxed{5}\ \boxed{a\frac{b}{c}}\ \boxed{6}\ \boxed{=}$$

to give an answer of $\frac{7}{12}$.

Alternatively, the calculation can be done without a calculator:

$$\frac{7}{10} \times \frac{5}{6} = \frac{35}{60}$$

which is $\frac{7}{12}$ in its lowest form.

(b) $\frac{8}{15} \div \frac{2}{5} = \frac{8}{15} \times \frac{5}{2} = \frac{40}{30} = \frac{4}{3} = 1\frac{1}{3}$

which could also have been obtained more directly from the key sequence:

$$\boxed{8}\ \boxed{a\frac{b}{c}}\ \boxed{15}\ \boxed{\div}\ \boxed{2}\ \boxed{a\frac{b}{c}}\ \boxed{5}\ \boxed{=}$$

(c) $\frac{3}{5} \times \frac{2}{3} \div \frac{4}{5} = \frac{3}{5} \times \frac{2}{3} \times \frac{5}{4} = \frac{1}{2}$, after cancellation.

Using the calculator

$$\boxed{3}\ \boxed{a\frac{b}{c}}\ \boxed{5}\ \boxed{\times}\ \boxed{2}\ \boxed{a\frac{b}{c}}\ \boxed{3}\ \boxed{\div}\ \boxed{4}\ \boxed{a\frac{b}{c}}\ \boxed{5}\ \boxed{=}$$

gives the same answer.

(d) $1\frac{3}{5} \times 1\frac{1}{2} = \frac{8}{5} \times \frac{3}{2} = \frac{24}{10} = 2\frac{2}{5}$, or

$$\boxed{1}\ \boxed{a\frac{b}{c}}\ \boxed{3}\ \boxed{a\frac{b}{c}}\ \boxed{5}\ \boxed{\times}\ \boxed{1}\ \boxed{a\frac{b}{c}}\ \boxed{1}\ \boxed{a\frac{b}{c}}\ \boxed{2}\ \boxed{=}$$

gives the same answer, $2\frac{2}{5}$.

(e) $3\frac{1}{8} \div 1\frac{1}{4} = \frac{25}{8} \div \frac{5}{4} = \frac{25}{8} \times \frac{4}{5} = \frac{100}{40} = \frac{5}{2} = 2\frac{1}{2}$.

Show that this answer is correct using the fraction button on your calculator.

Worked example

2.8 In his will John Brian left half of his personal estate to his widow, Gill Brian, and the remainder was split equally between his three children. If the value of his personal estate was £9600, determine how much was received by each member of his family.

Solution The widow received half of £9600, which is

$$\tfrac{1}{2}\left(\tfrac{9600}{1}\right) = £4800$$

For each child the fraction of the total personal estate is $\frac{1}{3}$ of $\frac{1}{2}$, or

$$\tfrac{1}{3}\left(\tfrac{1}{2}\right) = \tfrac{1}{6}$$

Each child therefore receives

$$\tfrac{1}{\cancel{6}_1}(\cancel{9600})^{1600} = £1600$$

Worked example

2.9 Brian Jackson wished to buy a house and needs to take out a mortgage to help purchase the house. A large bank will allow Mr Jackson to borrow $2\frac{3}{4}$ times his annual salary, whereas his local building society will only lend him $2\frac{1}{2}$ times his salary. If Mr Jackson's annual salary is £8400 determine the size of loan that each institution will give him.

Solution The mixed number, $2\frac{3}{4}$, is equal to the improper fraction $\frac{11}{4}$. (The denominator of the improper fraction is the same as the denominator of the fractional part of the whole number, and the numerator is obtained from $2(4) + 3 = 11$.)

From the bank Mr Jackson can borrow

$$2\tfrac{3}{4}(8400) = \tfrac{11}{\cancel{4}_1}(\cancel{8400})^{2100} = 11(2100) = £23\,100$$

From the building society he can borrow

$$2\tfrac{1}{2}(8400) = \tfrac{5}{\cancel{2}_1}(\cancel{8400})^{4200} = 5(4200) = £21\,000$$

Self-assessment question 2.2

1. Rewrite the following fractions with a denominator of 120 and then put the fractions in order of size: $\frac{3}{5}$, $\frac{1}{2}$, $\frac{5}{8}$, $\frac{7}{12}$

Exercise 2.2

1. The annual profit of a small business is £12 500. It is shared between two partners, Andrew and Brian. If Andrew receives $\frac{2}{5}$ of the profit, how much money does Brian receive?

2. Calculate
   (a) $\frac{3}{4} + \frac{2}{3}$ (b) $\frac{7}{8} - \frac{1}{6}$ (c) $1\frac{1}{3} + 1\frac{5}{6}$

3. An alloy consists of $\frac{7}{10}$ of copper, $\frac{9}{50}$ of tin and $\frac{3}{25}$ of zinc. How much of each metal is there in 750 grams of the alloy?

4. Calculate
   (a) $\frac{5}{6} \times \frac{2}{5}$ (b) $\frac{7}{8} \div \frac{1}{2}$ (c) $1\frac{2}{3} \times \frac{5}{6}$

5. A school has 1200 pupils, of whom $\frac{1}{6}$ are in the Upper School, $\frac{3}{8}$ are in the Middle School, and the rest are in the Lower School. How many pupils are there in the Lower School?

2.3   Decimal numbers

Decimals are fractions expressed in tenths, hundredths, thousandths, etc., and as normal counting takes place in tens, decimals are easier to work with than common fractions. To identify where the whole number ends and the fractional part begins, a decimal point is used. Thus 4.8 is equivalent to $4\frac{8}{10}$ and 0.27 is equivalent to $\frac{27}{100}$. When using a calculator decimal numbers are more useful than common fractions; for example, the answer to the calculation

$$\frac{11(3)}{8}$$

can be obtained by using the key sequence

The answer displayed is 4.125, a decimal number. Although many good calculators do have the facility to handle common fractions it is easier to use decimals. The answer, 4.125, to the above problem is equivalent to the common fraction

$$4.125 = 4 + \frac{1}{10} + \frac{2}{100} + \frac{5}{1000}$$
$$= 4\frac{125}{1000}$$
$$= 4\frac{1}{8}$$

in its lowest form.

Since the 1960s the United Kingdom has gradually undergone a process of 'decimalization'. This process has resulted in a much greater use of decimal numbers. It is quite common now to meet statements such as:

- the weight of a parcel is 1.35 kg
- interest rates are 11.875%
- a petrol tank holds 52.8 litres

whereas in previous years equivalent statements would have used:

- 2 lb 13 ounces
- $11\frac{7}{8}\%$
- 11 gallons 5 pints

As well as being more commonly used nowadays, decimal fractions have the advantage over common fractions that arithmetic is somewhat easier. For example, addition (and subtraction) of decimals is exactly the same as the addition (and subtraction) of whole numbers if a calculator is being used and there should be no problems. Try it!

$$8.32 + 16.1 + 0.2374 = 24.6574$$

If this calculation is carried out by hand it is important to keep the decimal points in a vertical line:

$$
\begin{array}{r}
8.32 \\
16.1 \\
\underline{0.2374} \\
24.6574
\end{array}
$$

Again multiplication of (or division between) two decimal numbers is similar to that between whole numbers, but care must be taken to ensure that the decimal point is in the correct position.

The answer to the problem

$$8.23 \times 6.1 = 50.203$$

has three digits to the right of the decimal point which is the same as the total number of digits to the right of the decimal point in the two numbers 8.23 and 6.1. However, be careful as some of the digits may be zeros. To avoid making very large errors it is safer to estimate the answer when multiplying or dividing decimals. For example, $8.91 \div 7.26$ is just a little more than 1, so if you get the answer to be 12.27 you would know this to be wrong (1.227 is the correct answer).

Worked example

2.10   Evaluate

(a) 27.4(6.3)   (b) 100(7.6547)   (c) 15.7 ÷ 4.15

Solution   (a) The answer will be above $25(6) = 150$, but below $30(7) = 210$. In fact,

$$27.4(6.3) = 172.62$$

(b) $100(7.6547) = 765.47$

It can be seen that multiplication of a decimal number by 100 simply moves the decimal point two places to the right. Similarly, the decimal point is moved one place to the right if the number is multiplied by 10, or it is moved one place to the left if it is divided by 10.

(c) The answer will be a little less than 4. Using a calculator

$$15.7 \div 4.15 = 3.783\,132\,53$$

The answer displayed has eight digits to the right of the decimal, which is typical of division involving decimal numbers. This degree of accuracy is not usually required and so the answer should be rounded to the degree of accuracy appropriate to the situation. For example,

$$15.7 \div 4.15 = 3.783$$

is correct to four significant figures (or, equivalently, correct to three decimal places).

Worked example

2.11   An electricity bill is made up of a standing charge plus a further charge per unit used. If the standing charge is £8.50 and the cost per unit is 5.32 pence, determine the total cost of the bill if

(a) 150 units,
(b) 450 units,

are used.

Solution   (a) The cost if 150 units are used is, in pence,

$$850 + 150(5.32) = 1648 \quad (or \; £16.48)$$

Beware! It is essential that the same units are used in the calculation. Do not mix pounds with pence.

(b) The cost if 450 units are used is

$$850 + 450(5.32) = 850 + 2394 = 3244 \; \text{pence} \quad (or \; £32.44)$$

Worked example

2.12 A school-leaver when looking for employment in a Job Centre notices three situations that appeal to her. The jobs and associated pay are shown in Table 2.2. In order to compare the pay, determine the weekly pay of each of the jobs.

Table 2.2.
Pay description of
three jobs

Job	Pay
Computer operator	£616.84 per month
Administrator	£8085.00 per annum
Sales assistant	£3.95 per hour (40 hour week)

Solution   Computer operator: There are 12 months in a year so the annual pay is

$$616.84 \times 12 = £7402.08$$

The weekly pay is equal to the annual salary divided by 52, giving

$$7402.08 \div 52 = £142.35$$

rounded to the nearest penny. You should note that it is not safe to assume that there are four weeks in a month; this would give a different, and incorrect, answer.

Administrator: The weekly pay is

$$8085 \div 52 = £155.48.$$

Sales assistant: The weekly pay is

$$3.95 \times 40 = £158.00$$

In terms of weekly pay the sales job is the most preferable.

Worked example

2.13 A lorry weighs 4.725 tonnes when empty. It is loaded with 37 machines each weighing 0.055 tonnes. What is the total weight of the lorry when loaded, correct to one decimal place?

Solution   The total weight of the machines is

$$37 \times 0.055 = 2.035 \text{ tonnes}$$

so the total weight of the lorry is

$$4.725 + 2.035 = 6.76 \text{ tonnes}$$

As this value is nearer 6.8 tonnes than 6.7 tonnes, the required answer is 6.8 tonnes.

Worked example

2.14   Express these numbers correct to two decimal places:

(a) 8.793   (b) 7.216   (c) 12.225

Solution   The answers are

(a) 8.79

(b) 7.22

(c) 12.225 is the same distance from 12.23 as from 12.22. In some respects both answers are equally correct, but convention is that we round up when there is this situation. So the given answer is 12.23.

Self-assessment questions 2.3

1. What do the decimal numbers 0.4 and 2.12 mean?

2. Write
   (a) 3.5 in fractional form
   (b) $3\frac{1}{4}$ in decimal form.

Exercise 2.3

1. Calculate
   (a) $7.29 \div 3.729 - 4.612$
   (b) $3.792 + 1.793 - 5.111$

2. Work out
   (a) 8.3×7.46
   (b) $8.2 \div 5.3$

3. Express the following numbers correct to one decimal place:
   (a) 8.439
   (b) 7.294

4. A consignment of nails has a total weight of 872 grams. If each nail weighs approximately 1.26 grams, estimate the number of nails in the consignment.

5. Which is greater, $\frac{7}{8}$ or 0.872?

Test and assignment exercises 2

1. The Olympic 800 metres for men was won with a time of 103.0 seconds in 1984 and with a time of 109.2 seconds in 1948. Calculate the difference in the two winning times.

2. How many pieces of wood each $5\frac{1}{2}$ centimetres long can be cut from a strip $123\frac{3}{4}$ centimetres long? Ignore the thickness of the saw-cut.

3. Evaluate the following expressions:

   (a) $\frac{3}{4} + \frac{5}{8}$ (b) $\frac{5}{9} - \frac{7}{18}$ (c) $\frac{4}{7}\left(\frac{3}{8}\right)$

4. On Monday John worked $6\frac{3}{4}$ hours, on Tuesday he worked $7\frac{1}{2}$ hours, on Wednesday $8\frac{5}{6}$ hours, on Thursday $7\frac{1}{4}$ hours, and on Friday, $5\frac{2}{3}$ hours. What total time did he work? Determine his pay for this week if he is paid at £3.22 per hour.

5. Evaluate the following expressions using decimals:
   (a) $9.421 - 5.3264$ (b) $5.71 \div 21.5$
   (c) $4.29(7.31 - 4.216)$

6. The circumference of a circle is given by $2\pi r$, where r is the radius of the circle and π is often given as $\frac{22}{7}$. Determine the circumference of the circle with radius $10\frac{1}{2}$ centimetres. Compare your answer with the answer obtained using the value of π given by your calculator.

7. A colour television is advertised at £384.95. An alternative method of purchasing the television is to make 24 payments of £17.19. How much extra do you pay by purchasing it on the instalment plan?

8. The national rate for the cost of sending parcels within the United Kingdom is given in Table 2.3. A company wishes to send eight parcels. The weight of each of these parcels is 0.85 kg, 1.57 kg, 1.90 kg, 2.74 kg, 4.61 kg, 5.75 kg, 6.40 kg and 9.40 kg. Calculate the total cost of sending these eight parcels.

Table 2.3. Inland postal rate

Weight not over	Rate (£)
1 kg	2.70
2 kg	3.35
4 kg	4.90
6 kg	5.50
8 kg	6.30
10 kg	7.30
30 kg	8.55

9. A litre bottle of whisky cost a publican £14.20. A measure of whisky is 25 ml and is sold at a price of 115 pence.
   (a) Determine the number of measures of whisky per bottle.
   (b) Determine the contribution to profit of selling all of these measures.
   (c) What fraction is the cost of the bottle compared with the money received?

10. A person spends $\frac{1}{4}$ of a 24-hour day asleep, $\frac{3}{8}$ at work, $\frac{1}{12}$ travelling, and the remainder is free time. What fraction of the day is free time?

11. A car achieves 42.7 miles per gallon under normal driving conditions. If the petrol tank holds 9.6 gallons, determine the expected travelling distance on one full tank of petrol.

12. An electrician needs copper wire in lengths of $3\frac{1}{8}$ metres. How many such lengths can be cut from a wire of length 250 metres?

3 Percentages and ratios

After reading this chapter, you should be able:

- to explain the meaning and usefulness of percentage and ratio
- to change a percentage to a fraction and a fraction to a percentage
- to express one number as a percentage of another number
- to compare percentages, fractions and decimals
- to use percentages in everyday situations
- to simplify ratios
- to express a ratio as either a fraction or a percentage
- to use ratios in everyday situations

3.1 Introduction

Percentages and ratios are used to indicate the relative size or proportion of the total rather than the absolute size. References to percentages and ratios are frequently seen in everyday situations. For example, we may see the following sign in a shop window:

SALE: 20% OFF

or, a national newspaper may have the following headline:

ANNUAL INFLATION UP TO 3.9%

In addition, a university may quote its student–staff ratio in the following form:

University SSR 18.5 : 1

People often meet percentages and ratios in everyday life when calculating values involving VAT, mortgage rates, discounts and commission.

Both percentages and ratios are used to indicate the relative size or proportion of one number compared with another. For example, a student may achieve a mark of 26 out of 40 in test A, and 28 out of 50 in test B. It is difficult, at the moment, to compare the relative performance of the student in the two tests as the total mark for each test was different. This chapter will identify ways of comparing the performances in the two tests.

3.2   Percentages

In life we are often interested in how numbers compare with one another. For example:

● How does the sales price of a garment compare with its cost price?
● How does an employee's wage in one year compare with his wage in a previous year?

If inflation is down to 3% why are prices still going up?

One way of comparing two numbers is to give the difference between them. On its own this information may not be useful:

● The sales price of the garment is £8 more than the cost price.
● The employee's wage is £5 per week more than last year.

Possibly a more useful way to receive this information might be:

● The sales price of garments is £150 for every £100 costs.
● The employee earns £103 for every £100 earned in the previous year.

Percentages provide us with a way of giving this information:

● There is a 50% markup on these garments.
● The employee's wage has increased by 3%.

Percentages are used to represent the proportion of one number compared with another. The word *percentage* comes from the Latin words '*per centum*' meaning 'by the hundred'. Basically, percentages can be considered to be fractions with a denominator of 100, where the numerator appears alone with % after it. For example $\frac{1}{2}$ can be written as $\frac{50}{100} = 50\%$. Similarly $\frac{4}{5} = \frac{80}{100} = 80\%$ and $1 = 100\%$ (or all).

KEY POINT

In order to change a fraction to a percentage, multiply by 100, which is equivalent to converting the fraction to decimal form and moving the decimal point two places to the right.

Worked example

3.1   Convert $\frac{7}{20}$ to percentage form.

Solution   Either

$$\frac{7}{20}(100)\% = 35\%$$

or

$$\frac{7}{20} = 0.35 = 35\%$$

KEY POINT

Conversely, a percentage is converted to a fraction by dividing by 100, or equivalently take the decimal form and move the decimal point two places to the left.

Worked example

3.2   Convert 62.5% to fraction form.

Solution $62.5\% = \frac{62.5}{100} = \frac{5}{8}$

or

$62.5\% = 0.625 = \frac{5}{8}$

Decimals or fractions bigger than one correspond to percentages greater than 100%. For example, $2\frac{1}{5} = 2.2 = 220\%$.

Worked example

3.3   A workforce can be classified into managerial, skilled and unskilled as in Table 3.1. Express the number of employees in each class as a percentage of the total workforce.

Table 3.1.
Company staff
profile

Class	Number of employees
Managerial	25
Skilled	75
Unskilled	150

Solution   The total workforce is $25 + 75 + 150 = 250$. The percentage of employees is:

- in the managerial class $= \frac{25}{250} = \frac{1}{10} = 10\%$,
- in the skilled class $= \frac{75}{250} = \frac{3}{10} = 30\%$, and
- in the unskilled class $= \frac{150}{250} = \frac{3}{5} = 60\%$.

Note that $10\% + 30\% + 60\% = 100\%$, as these categories include the whole workforce.

Worked example

3.4   A survey was carried out to investigate employment activities of the population. In all, 600 individuals were questioned, of whom 60% were male. The survey concludes that 90% of the males were in full-time regular employment, and 70% of the females were in full-time regular employment. Out of the 600 who were interviewed, calculate

(a) how many of the males worked full-time,
(b) how many of the females worked full-time.

Solution   60% of 600 = $\frac{60}{100}(600) = 360$.

Thus 360 males and 240 females were interviewed.

(a) The number of males in full-time regular employment is

90% of 360 = $\frac{90}{100}(360) = 324$

(b) The number of females in full-time regular employment is

70% of 240 = $\frac{70}{100}(240) = 168$

Percentages have many practical applications in everyday life. For example, in the car, double-glazing and insurance industries, salespersons are often paid on the basis of their success at selling, and receive a *commission* which is a percentage of their sales income. A further important application of percentages involves the idea of discount. In order to stimulate sales a retailer may decide to offer his goods at a price less than its normal price. A *discount* is the monetary amount by which the normal price is reduced expressed as a percentage of the normal price. In addition to these important applications, taxation and investments are also normally expressed in percentage form.

Worked example

3.5   How much money does a salesman earn on a £300 sale if his commission is 15%?

Solution   Salesman's commission is

15% of £300 = $\frac{15}{100}(300) = £45$

Worked example

3.6   An umbrella initially costs £8 but then has VAT added at 17.5%. The retailer then offers a discount of 25%. Determine the selling price of the umbrella. Does the order of computing the tax and discount influence the price of the umbrella?

Solution   The VAT on the umbrella is

17.5% of £8 = $\frac{17.5}{100}(8) = £1.40$

So the cost of the umbrella including VAT but before discount is

£8 + £1.40 = £9.40

The reduction in price due to the discount is

25% of £9.40 = $\frac{25}{100}(9.40)$ = £2.35

Hence the price of the umbrella is £9.40 − £2.35 = £7.05.

What would happen if the price after discount is found first and the tax is added? Would the answer be the same?
If the umbrella cost is £8, then the discount would be

25% of £8 = $\frac{25}{100}(8)$ = £2

giving a price after discount but before tax is added of £8 − £2 = £6. The tax can then be computed to be

17.5% of £6 = $\frac{17.5}{100}(6)$ = £1.05

When this is added to £6 it gives a selling price for the umbrella of £7.05, thus agreeing with the previous answer.

Under most conditions, if a problem involves two percentages it does not matter in which order the percentage calculations are performed.

Worked example

3.7 A woman invests £800 for two years. She has compound interest added at 5% per annum. How much interest is added over the two years?

Solution At the start of the two year period the investment is worth £800. During the first year the interest earned is

5% of £800 = $\frac{5}{100} \times$ £800 = £40

When this is added, the investment is worth £840 at the start of the second year. During the second year the interest earned is

5% of £840 = $\frac{5}{100} \times$ £840 = £42

Over the two years, the total interest earned by the investment is £40 + £42 = £82.
This worked example is an illustration of compound interest, which will be developed in Chapter 4.

Worked example

3.8 Brian has a number of deductions from his gross monthly wage. Nine per cent of this wage goes in National Insurance contributions, 6% goes in pension contributions, and 21% goes in tax. If his gross monthly wage is £400, determine the amount he pays towards National Insurance, his pension, and tax, together with his net monthly wage.

Solution

Gross monthly wage		£400
National Insurance $\frac{9}{100} \times$ £400	£36	
Pension contribution $\frac{6}{100} \times$ £400	£24	
Tax payment $\frac{21}{100} \times$ £400	£84	
Total contribution	£144	
Net monthly wage		£256

Self-assessment question 3.2

1. Rewrite the statements below using the % symbol:
   (a) In a quality control inspection 3 out of every 100 items were faulty.
   (b) In a university 38 students out of every 100 are female.
   (c) Eighteen out of every 100 students are left-handed.

Exercise 3.2

1. A pullover has been reduced by 25% in the sale. The original price was £28. How large is the reduction?

2. A woman invests £1000 in an account, offering compound interest at 10% per annum. How much will she have at the end of the second year?

3. A student gains 14 marks out of 40. Express this mark as a percentage.

4. Calculate
   (a) 30% of 80,
   (b) 45% of 120,
   (c) 60% of 45.

5. A debt of £600 is increased by 9%. What is the size of the debt now?

6. There are 18 empty houses in a village out of a total of 120 houses. What percentage of the total number of houses is empty?

7. Change the following fractions to percentages:
   (a) $\frac{3}{5}$ (b) $\frac{7}{8}$ (c) $\frac{13}{20}$

8. Four out of 16 peaches are bad. What percentage of peaches are bad?

9. The costs of the three departments in a company are

   Department A £24 000
   Department B £30 000
   Department C £66 000

   Find the percentage of company costs for each department

10. A meal costs £7.50, but then has VAT added at 17.5% and a service charge at 10%. Obtain the final cost of the meal.

3.3   Ratios

It is often of interest in everyday activities to compare two, or sometimes more, quantities; for example, in a work group there may be 12 apprentices together with 3 qualified technicians. In this situation we can say that the ratio of technicians to apprentices is 12:3. However, as in the case of fractions, it is more usual to express this in its lowest terms, or simplest form. Here 12 and 3 are both divisible by 3 so in its lowest terms the ratio 12:3 is four apprentices for each qualified technician. A specific feature of a ratio is that it gives no indication of size; it is merely used as a basis for comparison. In the above example a ratio of 4:1 is not only appropriate for 12 apprentices to 3 technicians but the same ratio could be used for 20 apprentices to 5 technicians or 40 apprentices to 10 technicians.

Suppose that a prospective student is considering attendance at one of two universities, Ayford University or Beeton University. The decision as to which of these two universities he attends is dependent on two ratios, the male-to-female ratio and the student–staff ratio. The relevant information is provided in Table 3.2.

Table 3.2.
Staff and student numbers

	Ayford University	Beeton University
Number of staff	150	200
Number of male students	1600	1800
Number of female students	800	1200

At Ayford University the ratio of male students to female students is

1600:800

or, in lowest terms,

2:1

At Beeton University the ratio of male students to female students is

1800:1200

or

3:2

(This might be put in the form 1.5:1 when making comparisons.)

Thus the ratio of males to females is greater at Ayford University; that is, there are more males per female at this university even though there are more males at Beeton University.

The total number of students at Ayford is 2400, giving a student–staff ratio of

2400:150

or

16:1

Similarly the student–staff ratio at Beeton is

3000 : 200

or

15 : 1

On the basis of this information we can expect slightly more personal tuition at Beeton as there the number of students per staff member is 15 compared with 16 at Ayford.

Worked example

3.9 An opinion poll on election party preferences was conducted, with replies received from 1000 interviewees. It was found that 480 favoured the Conservatives, 360 favoured the Labour Party, 120 preferred the Liberal Democrats, and the remaining 40 were undecided. Determine

(a) the ratio of decided to undecided,
(b) the ratio of Conservatives to Labour,
(c) the ratio of the two main parties to the rest of the population.

Solution (a) Altogether there are 960 interviewees with specified party preferences compared with 40 who do not specify a preference. The ratio of decided to undecided is 960 : 40, or 24 : 1.

(b) The ratio of Conservatives to Labour is 480 : 360 or, in lowest terms, 4 : 3.

(c) The number of interviewees who support the two main parties is 840, the remaining 160 either prefer the Liberal Democrats or are undecided. Hence the required ratio is 840 : 160 or 21 : 4 (or it might be given as 5.25 : 1).

Worked example

3.10 A chemical analysis on impurities in river water that passed a given location showed that in 100 parts of impurities, 60 were dangerous, 30 were harmless, and 10 were unidentified. What is the ratio of

(a) dangerous to harmless impurities,
(b) identified to unidentified impurities?

If river flow was such that 140 kilograms of impurities flowed past the given point each day, how many kilograms of each type would be present in the water each day?

Solution (a) The ratio of dangerous to harmless impurities is 60 : 30 or 2 : 1.

(b) The ratio of identified to unidentified impurities is 90 : 10 or 9 : 1.

Amount of dangerous impurities is $\frac{60}{100}$ of 140 kg, giving

$$\frac{60}{100}(140) = 84\,\text{kg}$$

Amount of harmless impurities is

$$\frac{30}{100}(140) = 42\,\text{kg}$$

leaving 14 kg that are unidentified.

Ratios are most frequently used in situations where the proportions of ingredients of a specified mixture remain the same no matter what the overall size of the total product. Consequently the amounts of the various ingredients of mixtures such as concrete, fertilizer, recipes and cocktails are often described in ratio form. For example, concrete consists of a mixture of cement, sand and gravel in the ratio two parts cement to three parts sand and five parts gravel.

Worked example

3.11 Given that concrete is made up of its three components cement, sand and gravel in the ratio $2:3:5$ determine how much of each component is needed to make 1500 kg of concrete.

Solution The total number of parts specified by the ratio is $2 + 3 + 5 = 10$. Thus $\frac{2}{10} = \frac{1}{5}$ of the mixture is cement, $\frac{3}{10}$ is sand, and $\frac{5}{10} = \frac{1}{2}$ is gravel.

$$\text{Amount of cement} = \tfrac{1}{5}(1500)\,\text{kg} = 300\,\text{kg}$$
$$\text{Amount of sand} = \tfrac{3}{10}(1500)\,\text{kg} = 450\,\text{kg}$$
$$\text{Amount of gravel} = \tfrac{1}{2}(1500)\,\text{kg} = 750\,\text{kg}$$

Worked example

3.12 A particular brand of fertilizer contains two parts nitrogen to one part potash and four parts phosphates by weight. If a specific quantity of this fertilizer contains 4 g of nitrogen, how much potash and phosphates will it contain?

Solution According to the ratio the weight of potash is half that of nitrogen, and the weight of phosphates is twice that of nitrogen. Hence

$$\text{weight of potash} = \tfrac{1}{2}(4\,\text{g}) = 2\,\text{g}$$

and

$$\text{weight of phosphates} = 2(4\,\text{g}) = 8\,\text{g}$$

Self-assessment question 3.3

1. Put each of the following ratios in the form ? : 1.
(a) 4000 : 1000
(b) 35 : 14
(c) 108 : 36

Exercise 3.3

1. Simplify the following ratios:
   (a) 4 : 8   (b) 5 : 30   (c) 36 : 24

2. Simplify the following ratios:
   (a) £4 : 50p   (b) 4 km : 1 cm
   (c) 20 g : 1 kg

3. Two metal bars have lengths 25 cm and 40 cm respectively. Give the ratio of the two lengths in its simplest form.

4. Two boys share £2 pocket money, with the younger boy receiving less than the older one. They share the money in the ratio 3 : 2. How much does the younger boy receive?

5. In a class of 60 students, the male : female ratio is 1 : 2. How many female students are there in the class?

6. A sum of money, £6000, is divided into three parts in the ratio 5 : 3 : 4. Determine the largest portion.

Test and assignment exercises 3

1. Convert each of the following to percentage form:
   (a) $\frac{1}{4}$ (b) 0.1   (c) 0.167

2. Convert each of the following percentages to decimals:
   (a) 54%   (b) 8.3%   (c) 40%

3. Find
   (a) 50% of 84   (b) 35% of 16
   (c) 120% of 45   (d) 3% of 700
   (e) 50% of 70% of 180

4. If 18 men from a shift of 300 are absent from work, what percentage is present?

5. How much money does a salesman earn on a £750 sale if his commission is 2%?

6. A television normally sells for £250 but is currently on sale at a discount of 20%. What is its sale price at the moment?

7. A couple go out for a meal at a restaurant. The initial cost of the meal is £15 but then VAT is added at 17.5% and a service charge is added at 10%. What is the total price for the meal?

8. A car costs £4000 new. If it depreciates in value by 12% during the first year, determine its value at the end of this year. If it depreciates by a further 12% during the second year, what is its value after two years? What is this final value as a percentage of the initial costs?

9. Put each of the following ratios in their lowest terms:
   (a) 3:6   (b) 360:48   (c) 68:17

10. A university has 3750 students and 125 staff. What is the student–staff ratio?

11. A father earns £200 per week and his son earns £120 per week. What is the ratio of their earnings?

12. Green paint is mixed from yellow and blue paint according to the ratio 7:3. Calculate how much of each colour is needed to make 50 litres of green paint.

13. A fertilizer contains the active ingredients nitrogen, potash and phosphorus in the ratio 2:1:1. If the fertilizer contains 5 g of potash, how much nitrogen and phosphorus does it contain?

14. At the last general election 34 000 people voted in a local constituency. If they voted for the Liberal Democrats, Conservative and Labour parties in the ratio 8:3:6, how many votes did each party receive?

15. The scale on a road map is given by 1 cm : 2 km. If two towns, Exford and Wyemouth, are 2.5 cm apart on the map, what is the actual distance between these towns?
    Two further towns are known to be 17.5 km apart; what would be the distance between the two towns on the road map?

16. A student achieves a mark of 26 out of 40 in test A, and 28 out of 50 in test B. In which of the two tests did the student achieve the higher percentage score?

4 Powers and roots

<table>
<tr><td>**Objectives**</td><td>After reading this chapter, you should be able:

• to evaluate expressions involving powers and roots

• to solve real problems requiring powers using a calculator

• to use scientific notation to represent very large and very small numbers</td></tr>
</table>

4.1 Introduction

In the three previous chapters a number of examples involving only addition, subtraction, multiplication and division of numbers were described. More difficult calculations are required time and time again in finance and can be made more simple by using powers and roots. Indeed, you would not be able to solve such financial problems without a knowledge of powers and roots.

4.2 Powers

A multiple-choice test is one in which the respondent has a choice of answers for each question. A typical example of such a test is shown in Table 4.1. Try this test; it should not cause too many problems.

What letters have you put in the five boxes on the right-hand side of the test paper? You should have the letters C, B, A, D, D in this order. If you do not get these answers, try the question(s) again or read the appropriate chapter of this text.

In question 1 of the multiple-choice test there are four possible answers even though only one is correct, which is also true in question 2. How many possible arrangements of answers are there for these two questions? For example, one possible arrangement is AA (A for question 1 and A for question 2) even though both answers are incorrect, or AB, etc. It should not take long to check that there are 16 possible arrangements:

Table 4.1. Numeracy multiple- choice test	*Arithmetic test* For each question one of the answers is correct. Write A, B, C or D in the box to the right of the question.

Question 1 ☐

Evaluate $5 - 3(4) + 5$

A: 22   B: 13
C: -2 D: 18

Question 2 ☐

Evaluate $\frac{3}{19}\left(\frac{2}{9} + \frac{5}{6}\right)$

A: 6.92   B: $\frac{1}{6}$

C: $\frac{1}{3}$ D: $\frac{6}{9}$

Question 3 ☐

Calculate $2.531(4.2) - 3.707$ correct to 2 decimal places

A: 6.92   B: 1.248
C: 6.923   D: 6.93

Question 4 ☐

Find 15% of £23.60

A: £15.73   B: £3.45
C: £1.57   D: £3.54

Question 5 ☐

A camera normally selling for £102.40 is on sale at a
discount of 25%. What is its sale price?

A: £25.60   B: £78.60
C: £75.50   D: £76.80

AA, AB, AC, AD, BA, BB, BC, BD, CA, CB, CC, CD, DA, DB,
DC, DD

An alternative method of determining the possible number of
arrangements of answers to these two questions is to calculate

$$4(4) = 16$$

since there are four ways of selecting an answer to each of the two
questions. Similarly, if all five questions on the test are considered, the
total number of possible answers is

$$4(4)(4)(4)$$

which can be computed on a calculator using the key sequence

| 4 | × | 4 | × | 4 | × | 4 | × | 4 | = |

and gives the answer 1024. A shorthand way of writing this is 4^5, where 4 is the *base* and 5 is the *power,* and is called 'four raised to the power 5'. Therefore

$$4^5 = 4(4)(4)(4)(4) = 1024$$

If your calculator possesses an $\boxed{x^y}$ key or $\boxed{y^x}$ key, then 4^5 can be computed directly using the key sequence

$$\boxed{4} \quad \boxed{x^y} \quad \boxed{5} \quad \boxed{=}$$

Similarly,

$$3^4 = 3(3)(3)(3) = 81$$
$$2^6 = 2(2)(2)(2)(2)(2) = 64$$

can be found directly using the $\boxed{x^y}$ button on your calculator.

Note that when the base is ten, the product is easy to find:

$$10^2 = 10(10) = 100$$
$$10^3 = 10(10)(10) = 1000$$
$$10^4 = 10(10)(10)(10) = 10\,000$$

because the power is always exactly equal to the number of zeros in the answer. It follows that $10^1 = 10$ and $10^0 = 1$. In general any number to the power of 1 is equal to itself and any number to the power of 0 is equal to 1. Thus

$$2^1 = 2 \qquad 2^0 = 1$$

$$5^1 = 5 \qquad 5^0 = 1$$

KEY POINT

> Any number raised to the power 0 equals 1;
>
> that is, $a^0 = 1$
>
> Any number raised to the power 1 equals the number;
>
> that is, $a^1 = a$

Worked example

4.1 Use the $\boxed{x^y}$ key on your calculator to evaluate

(a) 6^3 (b) 8^1 (c) $(-2)^5$
(d) 7^0 (e) $(3^2)(3^3)$ (f) $4^5 \div 4^2$

Solution (a) $6^3 = 216$
(b) $8^1 = 8$
(c) $(-2)^5 = -32$

Some calculators display an error message when using the $\boxed{x^y}$ key for negative values of x. If this is the case evaluate 2^5 and change the sign of the answer if the power is odd, as is the case here.

(d) $7^0 = 1$

(e) There are two ways of getting the correct answer here:

$$(3^2)(3^3) = (9)(27) = 243$$

Alternatively

$$(3^2)(3^3) = 3(3)3(3)(3) = 3^5 = 243$$

(f) Again there are two methods of solution:

$$4^5 \div 4^2 = 1024 \div 16 = 64$$

or

$$4^5 \div 4^2 = \frac{4(4)(4)(4)(4)}{4(4)} = 4^3 = 64$$

The last two parts of this example illustrate an important property of powers, namely, that if the bases are the same, multiplication is achieved by adding the powers:

$$3(3^3) = 3^1(3^3) = 3^{1+3} = 3^4 = 81$$
$$2^3(2^4) = 2^{3+4} = 2^7 = 128$$

Similarly, provided the bases are the same, division is achieved by subtracting the powers:

$$8^3 \div 8 = 8^3 \div 8^1 = 8^{3-1} = 8^2 = 64$$
$$6^5 \div 6^2 = 6^{5-2} = 6^3 = 216$$

KEY POINT

> Providing the bases are the same, multiplication involves adding the powers, and division involves subtracting the powers.
>
> For example, $a^m \times a^n = a^{(m+n)}$ and $a^m \div a^n = a^{(m-n)}$

Do not forget that the bases must be the same for this property to apply. You cannot calculate $2^3 \times 3^4$ in this way.

Worked example

4.2   Evaluate $4^4 \times 2^4$.

Solution $4^4 \times 2^4 = 4(4)(4)(4) \times 2(2)(2)(2) = 256 \times 16 = 4096$

Note that $2^4 = 2(2) \times 2(2) = 4^2$, so the calculation may have been carried out in the following way:

$$4^4 \times 2^4 = 4^4 \times 4^2 = 4^6 = 4096$$

Using the division rule

$$3^3 \div 3^5 = 3^{3-5} = 3^{-2}$$

but what does 3^{-2} mean?

Remember $3^3 \div 3^5$ can also be computed from first principles:

$$\frac{3^3}{3^5} = \frac{3(3)(3)}{3(3)(3)(3)(3)} = \frac{1}{3^2}$$

This provides a definition of 3^{-2}, namely

$$3^{-2} = \frac{1}{3^2} = \frac{1}{3(3)} = \frac{1}{9}$$

Similarly

$$7^{-3} = \frac{1}{7^3} = \frac{1}{7(7)(7)} = \frac{1}{343}$$

$$5^{-1} = \frac{1}{5^1} = \frac{1}{5}$$

All of these answers can be obtained directly using the $\boxed{x^y}$ key on a calculator, although the answers are given in decimal form. For example, 5^{-1} is evaluated using the key sequence

$$\boxed{5} \quad \boxed{x^y} \quad \boxed{1} \quad \boxed{+/-} \quad \boxed{=}$$

and gives the answer 0.2.

Worked example

4.3 Write down the following in fraction form:

(a) 2^{-2} (b) 6^{-1} (c) $4^3 \times 4^{-5}$

Solution (a) $2^{-2} = \frac{1}{4}$

(b) $6^{-1} = \frac{1}{6}$

(c) $4^3 \times 4^{-5} = 4^{-2} = \frac{1}{4 \times 4} = \frac{1}{16}$

Self-assessment question 4.2

1. Explain how negative powers (e.g. 3^{-2}) are interpreted.

Exercise 4.2

1. Evaluate

 (a) 4^3 (b) 7^2 (c) 3^{-2} (d) 10^4

2. Work out

 (a) $4^2 \times 4^3$ (b) $3^2 \times 3^{-3}$ (c) $5^2 \times 3^3$

3. Compute

 (a) $10^2 \times 3^3$ (b) $3^2 \times 5^0$ (c) $2^3 \times 3^1$

4. Work out

 $5^3 - 11^2$

4.3   Roots

The inverse of raising to a power is finding the *root* of a number. For example,

- $\sqrt{49}$ = square root of 49 = 7, since $7^2 = 49$;
- $\sqrt[3]{64}$ = cube root of 64 = 4, since $4^3 = 64$;
- $\sqrt[4]{81}$ = fourth root of 81 = 3, since $3^4 = 81$.

Is this a square root?

Use the $\boxed{x^y}$ key on your calculator to evaluate $49^{\frac{1}{2}}$, $64^{\frac{1}{3}}$, $81^{\frac{1}{4}}$.
The answers are, as above, 7, 4 and 3. For example,

$$\boxed{81}\ \ \boxed{x^y}\ \ \boxed{0.25}\ \ \boxed{=}$$

gives 3, but note that the fractional powers $\frac{1}{2}$, $\frac{1}{3}$, $\frac{1}{4}$ must be input as decimals. To overcome this problem use the $\boxed{x^{1/y}}$ key to evaluate roots.

Worked example

4.4 Use your calculator to evaluate the following expressions:

(a) $4^2(4^4)$ (b) $2^3(2^{-2})$ (c) $3^3 \div 3^6$

(d) $\sqrt{30}$ (e) $\sqrt{-16}$ (f) $2^{\frac{3}{2}}(2^{\frac{5}{2}})$

(g) $1.57^{\frac{1}{3}}$ (h) 0.8^{-2}

Solution

(a) $4^2(4^4) = 4^6 = 4096$

(b) $2^3(2^{-2}) = 2^1 = 2$

(c) $3^3 \div 3^6 = 3^{-3} = \frac{1}{3^3} = \frac{1}{27}$

(d) Each of the following three key sequences will give the answer:

(i) $\boxed{30}\ \ \boxed{\sqrt{}}$

(ii) $\boxed{30}\ \ \boxed{x^{\frac{1}{y}}}\ \ \boxed{2}\ \ \boxed{=}$

(iii) $\boxed{30}\ \ \boxed{x^y}\ \ \boxed{0.5}\ \ \boxed{=}$

The given answer in each case is 5.4772

(e) The square root of a negative number does not exist. (The calculator displays an error message.)

(f) $2^{\frac{3}{2}}(2^{\frac{5}{2}}) = (2^{\frac{3}{2}+\frac{5}{2}}) = 2^4 = 16$

The addition property of powers applies even if the powers are fractions.

(g) Using the key sequence

$$\boxed{1.57}\ \ \boxed{x^{\frac{1}{y}}}\ \ \boxed{3}\ \ \boxed{=}$$

the answer is displayed as 1.162, to four significant figures.

(h) The key sequence

$$\boxed{0.8}\ \ \boxed{x^y}\ \ \boxed{2}\ \ \boxed{+/-}\ \ \boxed{=}$$

displays the answer 1.5625.

Note that in (d) (i) the key sequence for $\sqrt{30}$ was given as $\boxed{30}$ $\boxed{\sqrt{}}$. However, there are many different makes of calculator, and it is becoming more frequent for calculators to require the key sequence $\boxed{\sqrt{}}$ $\boxed{30}$. You must find out which way to use the keys on your calculator.

Worked example

4.5 Write down answers to:

(a) $\sqrt{169}$ (b) $512^{1/3}$ (c) $3.527^{1/2}$ to one decimal place

Solution   (a) 13, using the key sequence $\boxed{169}$ $\boxed{\sqrt{}}$.

(b) 8, using the key sequence $\boxed{512}$ $\boxed{x^{1/y}}$ $\boxed{3}$ $\boxed{=}$.

(c) 1.9 which has been rounded from 1.8780.

Self-assessment question 4.3

1. Explain how fractional powers (e.g. $9^{1/2}$) are interpreted.

Exercise 4.3

1. Write down the square roots of the following numbers:
   (a) 144   (b) 484   (c) 25

2. Find
   (a) $\sqrt{37\,000}$ to two decimal places
   (b) $\sqrt{0.0125}$ to three decimal places

3. Find
   (a) $\sqrt{4/9}$ (b) $\sqrt{25/49}$

4. Write down
   (a) $64^{1/3}$ (b) $216^{1/3}$

4.4   **Practical examples of the use of powers and roots**

A number of practical problems involving money can only be solved using powers of numbers. An example of this is the calculation of compound interest when money has been invested in a bank account or savings account. The following result applies when interest is accumulated in a compounded way.

| KEY POINT | Compound Interest Formula |

If an amount P is invested at an interest rate of $r\%$ per annum, the resulting amount after n years is given by

$$P\left(1 + \tfrac{r}{100}\right)^n$$

Worked example

4.6 Use the compound interest formula to determine the value of the following two investments:

(a) £500 invested at 11% p.a. for three years,
(b) £2000 invested at 9% p.a. for six months.

Solution (a) $P = 500$, $r = 11$, $n = 3$

$$\text{Value} = 500\left(1 + \tfrac{11}{100}\right)^3$$
$$= 500(1.11)^3$$
$$= 500(1.367\,631)$$
$$= £683.82 \text{ (to the nearest penny)}$$

(b) $P = 2000$, $r = 9$, $n = \tfrac{1}{2}$

$$\text{Value} = 2000\left(1 + \tfrac{9}{100}\right)^{\frac{1}{2}}$$
$$= 2000(1.09)^{\frac{1}{2}}$$
$$= 2000(1.044\,03)$$
$$= £2088.06 \text{ (to the nearest penny)}$$

In Great Britain most families require a long-term loan (or mortgage) to purchase their home. Mortgages are usually repaid in monthly instalments, each of which includes interest on the unpaid balance and a payment on the principal (the amount borrowed).

| KEY POINT | Loan Repayment Formula |

The size of the annual repayments is given by the formula

$$\text{Annual repayment} = \frac{P(r)}{100} \frac{(1 + r/100)^n}{[(1 + r/100)^n - 1]}$$

where P = principal (size of loan),

$\quad\quad r$ = net interest rate (%) per annum,

$\quad\quad n$ = term of repayment in years.

The monthly repayment is then calculated by dividing the annual repayment by 12.

Worked example

4.7 Determine the monthly repayments for each of the loans listed in Table 4.2. (Note that the usual interest rate quoted on mortgage loans is the gross interest rate but as such loans are often liable to tax relief the net interest rate may be the one used in repayment calculations.)

Table 4.2.
Mortgage descriptions

Name	Loan (£)	Term (years)	Net interest rate (%)
Mr Arrowsmith	45 000	20	9.5
Mrs Brown	15 000	30	8.0
Mr Cheadle	25 000	25	7.5

Solution For Mr Arrowsmith,

$$P = 45\,000, r = 9.5, n = 20.$$

$$\left(1 + \tfrac{r}{100}\right)^n = \left(1 + \tfrac{9.5}{100}\right)^{20} = 1.095^{20} = 6.141\,612\,104$$

(It is important to keep as much accuracy as possible in the intermediate steps of a calculation. Indeed, you could store this number in the memory of your calculator to save keying it in again.)

$$\text{Annual repayment} = \tfrac{45\,000(9.5)}{100}\left(\tfrac{6.141\,612\,104}{5.141\,612\,104}\right)$$

$$= £5106.45$$

Mr Arrowsmith's monthly repayment $= \tfrac{5106.45}{12} = £425.54$

For Mrs Brown, $P = 15\,000$, $r = 8$, $n = 30$, giving an annual repayment of

$$\tfrac{15\,000(8)}{100}\left(\tfrac{1.08^{30}}{1.08^{30}-1}\right)$$

$$= 1200\left(\tfrac{10.062\,656\,89}{9.062\,656\,89}\right) = £1332.41$$

Her monthly repayments are $\tfrac{1332.41}{12} = £111.03$.

The annual repayment for Mr Cheadle is

$$\tfrac{25\,000(7.5)}{100}\left(\tfrac{1.075^{25}}{1.075^{25}-1}\right) = £2242.76$$

His monthly repayments are £186.90.

Self-assessment question 4.4

1. (a) John invests £8000 in an account earning 5% p.a. for three years. How much is this worth at the end of this time?
   (b) John also borrows £10 000 at 9% p.a. and makes monthly repayments over 15 years. Find the size of these repayments.

Exercise 4.4

1. Mr Patel invests £12 500 in an account earning 6.3% per annum for four years. How much interest accumulates over the period? Use the compound interest formula.

2. Find the value of the investment, £2500, invested at 6.5% for five years.

3. Mr Dixon borrows £50 000 over 25 years, and is charged interest at 8.5% per annum. Estimate his monthly repayments required to pay off the loan and the total amount paid over the 25 years.

4. Mrs Evans borrows £20 000 at 8.75% and agrees to make monthly repayments over 10 years. Find the size of the repayments.

4.5   Scientific notation

Finally in this chapter, properties of numbers written in scientific notation are investigated.

KEY POINT

A number is said to be written in *scientific notation* if it is expressed as a number between 1 and 10 together with some power of 10.

For example, 250 can be written as $2.5(10^2)$ or 0.36 can be written as $3.6(10^{-1})$. The main advantage of this notation is that very large or very small numbers can be written more compactly:

$$486\,000 = 4.86(10^5)$$

Most calculators use this notation when giving answers that are very large or small. Use your calculator to evaluate

$$0.000\,0023(0.000\,06)$$

It is likely that your calculator displays $1.38\ -10$, indicating that the answer is 1.38×10^{-10}, the power of 10 moving the decimal point 10 places to the left:

$$1.38 \times 10^{-10} = 0.000\,000\,000\,138$$

Now check the calculation

$$83\,000(165\,400) = 1.372\,82(10^{10})$$
$$= 13\,728\,200\,000$$

Worked example

4.8   Obtain $3.8 \times 10^3 + 9.72 \times 10^2$

Solution   This question can be carried out in a number of ways:

Firstly,

$$3.8 \times 10^3 + 9.72 \times 10^2 = 3800 + 972 = 4772 = 4.772 \times 10^3$$

Alternatively,

$$3.8 \times 10^3 + 9.72 \times 10^2 = 3.8 \times 10^3 + 0.972 \times 10^3$$
$$= (3.8 + 0.972) \times 10^3 = 4.772 \times 10^3$$

Self-assessment question 4.5

1. Express the following numbers in scientific notation:
   (a) 96 742   (b) 0.000 086   (c) 0.0090

Exercise 4.4

1. Convert the following scientific notation to normal format:
   (a) 3.75×10^5 (b) 9.87×10^{-4}
   (c) 7.52×10^{-3}

2. Subtract 7.5×10^{-3} from 8.75×10^{-1} and give the answer in scientific notation.

3. Obtain $6.25 \times 10^{-4} + 5.2 \times 10^{-2}$.

Test and assignment exercise 4

1. Evaluate
   (a) 8^3 (b) 9^0 (c) $(-2)^4$
   (d) 10^4 (e) 7^1

2. Use your calculator to evaluate the following expressions.
   (a) $4(4^2)$ (b) $6^3 \div 6^5$ (c) $7^2(7^{-3})$
   (d) $(-3)^{-2}$ (e) $3^{\frac{1}{2}}(3^{2\frac{1}{2}})$

3. A multiple-choice test, comprising 10 questions, allows five possible answers for each question. Determine the total number of arrangements of possible answers for this test.

4. Evaluate the following expressions
   (a) $\sqrt{64}$ (b) $\sqrt{40}$ (c) $\sqrt{-25}$
   (d) $12.6^{0.5}$ (e) $7.98^{\frac{1}{3}}$

5. An insurance company purchased a number of securities valued at £10 000 that guaranteed an interest rate of 8% per annum. If these securities were kept for three years, how much interest should they earn?

6. Evaluate
   (a) 0.01^2 (b) $\sqrt{0.01}$ (c) 0.01^{-1}

7. £750 was invested in a bank account at 9.5% per annum. Determine the value of this investment after 18 months.

8. A family takes out a mortgage over 25 years for £50 000. If the net interest rate is 8.4% calculate their monthly repayments and the total interest paid.

9. Express the following numbers in ordinary decimal notation:
   (a) $5.423(10^4)$ (b) $7.29(10^1)$
   (c) $8.61(10^{-2})$

10. Subtract $9.324(10^{-2})$ from $4.75(10^1)$ and give the answer in scientific notation.

Section B
Pictures

5 Tables

Objectives	After reading this chapter, you should be able:
	• to choose suitable class boundaries for making a frequency table
	• to arrange data into a frequency table
	• to extract appropriate information from tables

5.1 Introduction

It is essential that today's citizen can understand a set of tabulated data. So much information, from train timetables to holiday costs, is presented in this form. There are also occasions when it is necessary to condense a large set of data into a simple frequency table for easier understanding and analysis.

5.2 Frequency tables

Generally the collection of information from a survey or experiment leads to a large mass of data. In their original form, as a large set of numbers, these data appear almost meaningless. They need to be condensed into a more understandable form. The figures in Table 5.1 represent the number of matches inside a sample of 50 matchboxes.

The first step towards making it more presentable is to convert it to a *grouped frequency table*. This involves dividing the range of the

Table 5.1.
Number of matches per matchbox

232	251	237	255	229	246	244	249	241	235
250	248	245	246	234	239	250	242	240	246
238	249	247	243	236	250	248	244	247	239
252	247	246	241	244	245	245	243	237	246
254	252	245	243	242	244	246	242	239	244

data into classes and counting the number of items that fall into each class. The data here range from 229 to 255 and suitable classes are as shown in Table 5.2.

Table 5.2.
Suitable classes

Number of matches per box
229–231
232–234
235–237
238–240
241–243
244–246
247–249
250–252
253–255

There may be 247 but they are all used.

There are obviously many different ways in which the range 229 to 255 can be divided. In the classification given in Table 5.2 there are nine classes, all of equal size. There are no firm rules about how many classes to use but it is usual to have between 5 and 15 classes. Less than 5 would create a loss in information, whereas more than 15 would be too many to take in easily and so would rob the frequency table of its main reason for existence. In addition, if the classes are of equal size it gives a clearer view of the distribution of the data. 229–231 and 231–234 would be wrong because 231 would have two classes to belong to, and 229–231 and 233–235 are wrong because 232 would have no class to belong to.

KEY POINT

> The class boundaries must cover all the data with no gaps and have no ambiguities so that each data value has one and only one class to belong to.

Having set up the classes the next step is to count the number of data items falling into each. The usual way is to employ a tallying procedure. This involves taking each item in turn in the original data and entering a tally-mark against the appropriate class. It is a good idea to record the tally-marks using a 'five-bar gate' method so that they can be easily added up at the end. The number of tally-marks recorded against a class is the class frequency. The above example produces the grouped frequency table shown in Table 5.3.

Table 5.3.
Frequency table for the number of matches per box

Class	Tally marks	Frequencies
229–231	1	1
232–234	11	2
235–237	1111	4
238–240	⊞HT	5
241–243	⊞HT 111	8
244–246	⊞HT ⊞HT ⊞HT	15
247–249	⊞HT 11	7
250–252	⊞HT 1	6
253–255	11	2

Worked example

5.1   Table 5.4 shows a list of marks (out of 100) gained by 60 students in an examination. Arrange them in a grouped frequency table.

Table 5.4.
Examination marks

78	15	59	66	22	41	44	12	84	32
56	81	50	45	82	58	75	34	69	61
67	55	52	73	31	48	41	56	91	56
84	66	56	44	39	45	51	68	70	64
66	27	47	78	24	53	32	24	72	53
59	42	78	64	76	38	87	41	54	36

Solution   The smallest value is 12 and the largest is 91 so class intervals of 10–19, 20–29, etc. up to 90–99 will cover all the data. Using the tally system produces the frequency table shown in Table 5.5.

Table 5.5.
Frequency table for examination marks

Class	Tally	Frequency
10–19	11	2
20–29	1111	4
30–39	⊞HT 11	7
40–49	⊞HT ⊞HT	10
50–59	⊞HT ⊞HT 1111	14
60–69	⊞HT 1111	9
70–79	⊞HT 111	8
80–89	⊞HT	5
90–99	1	1

Self-assessment questions 5.2

1. Arrange the following 20 home team scores from recent Premier Division football matches into a frequency table.

    0, 1, 3, 0, 0, 1, 2, 0, 1, 1, 0, 0, 4, 1, 2, 0, 0, 1, 0, 2.

2. The following class intervals are proposed for a frequency table for the results from a survey on costs of meals in hotels. Criticize these suggested intervals.

    £0   to £3.00
    £3.00   to £5.00
    £5.00   to £9.00
    £10.00 to £15.00
    £15.00 and over

Exercise 5.2

1. Rework worked example 5.1 but use class intervals that are 10–29, 30–49, etc. Which class interval do you think is the more appropriate?

2. Using suitable class intervals form a frequency table from the following 30 house prices taken from a newspaper (all prices in £s):

33 500	124 000	95 250	47 000	66 750
155 950	206 100	37 350	29 450	41 490
88 750	64 850	39 370	44 380	110 500
36 450	52 690	58 660	108 850	28 900
50 750	84 690	44 750	115 500	73 590
66 750	30 500	41 950	54 500	63 750

3. A company manufactures components for use in the production of motor vehicles. The number of components produced each day over a 40-day period is tabulated in Table 5.6. Group the data into seven classes.

Table 5.6. Number of components produced over a 40-day period

353	328	326	305	362
338	344	346	346	328
317	332	349	349	344
355	323	345	309	332
332	325	324	345	323
333	336	319	324	325
335	335	331	319	336
349	317	318	331	365

5.3   Data tables

The previous examples give a good indication of how to construct a table from statistical data; however, in daily life it is more likely that an individual will have to read various types of table and interpret the information they contain. Many tables are straightforward to understand but some are much more complicated. The rest of this chapter looks at some common examples of tabulated information and in each case the information that is relevant is extracted. The tables in these examples display some or all of the features that well-presented tables must have to give clear and unambiguous information.

KEY POINT

All tables should have:

- a title
- clearly stated units
- a source, if appropriate
- footnotes to help clarification

Worked example

5.2   A well-known international car rental firm issued information shown in Table 5.7 showing its rent-a-car rates for 1997. Determine the appropriate total cost of

(a) hiring a Peugeot 206 (economy car) for one week with Collision Damage Insurance (CDI),

(b) hiring a Mondeo (standard size) for four days with CDI.

(c) If the Mondeo from part (b) is returned two hours late what will be the new cost?

Table 5.7.
Car hire charges

Code	Car type	Daily (£)	Weekly (£)
ECMN	Economy car up to 1100cc	23.25	148.75
CCMN	Mid range up to 1400cc	24.25	157.50
ICMN	Intermediate up to 1800cc	27.00	175.00
IWMN	Estate car up to 1800cc	34.00	220.50
SCMN	Standard size up to 2000cc	36.00	234.50
FCAN	Full size auto. up to 2000cc	43.00	280.00
LCAR	Luxury car automatic	61.00	395.50
Code	Commercial type	Daily (£)	Weekly (£)
11 AV	Escort type up to 10 cwt diesel van	25.00	125.00
12 BV	Transit/Merc. up to 18 cwt diesel van	30.00	150.00
13 CV	Transit 190 LWB Hi-Roof 35 cwt diesel van	40.00	200.00
14 TA	3.5 tonne tipper/dropside diesel	45.00	225.00
15 TB	3.5 tonne Luton box van with tail lift	45.00	225.00
16 TC	7.5 tonne dropside truck	65.00	325.00
17 TD	7.5 tonne curtainsider truck	70.00	350.00

Code	Commercial type	Daily (£)	Weekly (£)
18 TE	7.5 tonne box with tail lift truck	75.00	375.00
19 TF	7.5 tonne Tipper/HIAB truck	75.00	375.00

Rates

Rates include unlimited mileage and are inclusive of Value Added Tax and petrol charges. These rates are subject to change without notice.

Terms of rental

All rentals are subject to the full conditions which are detailed on the reverse of the Rental Agreement.

Delivery and collection

A free delivery and collection service is available within 5 miles of all UK offices to business premises, thereafter each mile will be charged at 75 pence per mile round trip. Vehicle delivered prior day to rental for early starts and over a weekend for Monday morning will be surcharged at £15.00.

Hourly charges

A charge of equivalent to one-third of the daily rate will be applied to all additional hours after the minimum rental period of 24 hours has been exceeded. After three hours, a full day's rate will be applied.

Insurance

Collision Damage Insurance (CDI)

ECMN–ICMN	£7.05	AV	£7.05
IWMN–FCAN	£8.22	BV–CV	£8.22
LCAR	£9.40	TA–TF	£9.40

Foreign Licences £9.40

CDI Accepted
A non waiverable insurance excess of £100 applies.

CDI Declined (renter's financial responsibility)

ECMN–ICMN	£2000	AV	£2000
IWMN–FCAN	£3000	BV–CV	£3000
LCAR	£4000	TA–TF	£4000
Foreign Licences	£4000		

Solution

(a) 1 week   £148.75
CDI   £7.05
Hence total cost is   £155.80

(b) 4 days' costs   4(£36.00) = £144.00
CDI   £8.22
Hence total cost is   £155.22

(c) Charge for each hour late is 1/3 of £36.00 = £12.00 and so the extra cost for two hours late is £24.00
Total cost is now £161.62 + £24.00 = £185.62

Worked example

5.3 Table 5.8 shows the numbers of marriages of various types for five different years in the past.

(a) How many marriages between two people who had both been divorced were there in 1991?

(b) What can be said about the number of first marriages for both partners between 1961 and 1992?

(c) Explain where the result 9% for 'remarriages of the divorced as a percentage of all marriages' was calculated in 1961.

Table 5.8. Marriages by type (UK, thousands)		1961	1971	1981	1991	1992
	First marriage for both partners	340	369	263	222	222
	First marriage for one partner only					
	Bachelor/divorced woman	11	21	32	32	35
	Bachelor/widow	5	4	3	2	2
	Spinster/divorced man	12	24	36	35	36
	Spinster/widower	8	5	3	2	2
	Second (or subsequent) marriage for both partners					
	Both divorced	5	17	44	45	47
	Both widowed	10	10	7	4	4
	Divorced man/widow	3	4	5	4	4
	Divorced woman/widower	3	5	5	4	4
	All marriages	397	459	398	350	356
	Remarriages[1] as a percentage of all marriages	14	20	34	36	38
	Remarriages[1] of the divorced as a percentage of all marriages	9	15	31	34	35

[1]Remarriage for one or both partners.
Sources: Office of Population Censuses and Surveys; General Register Office (Scotland); General Register Office (Northern Ireland). *Social Trends* 1993. Office for National Statistics. Crown Copyright, 1997.

Solution

(a) 45 000

(b) There is a clear decline over time from 340 000 in 1961 to 222 000 in 1992.

(c) Total remarriages of the divorced in 1961 is given by

Bachelor/divorced woman	11 000
Spinster/divorced man	12 000
Both divorced	5 000
Divorced man/widow	3 000
Divorced woman/widower	3 000
Total	34 000

Since in 1961 there were 397 000 marriages in total the required percentage is

$$(34\,000/397\,000)100 = 8.56 \text{ or } 9\%$$

Worked example

5.4 Table 5.9 can be found in many mathematical texts. Use this table to estimate, correct to three significant figures,

(a) $\sqrt{70}$

(b) $\sqrt{72.5}$

(c) n such that $1/n = 0.014$

Table 5.9.
Squares, cubes, square
roots and reciprocals

n	n^2	n^3	$\sqrt{n}$	$\sqrt[3]{n}$	$\sqrt{10n}$	$\sqrt[3]{10n}$	$\sqrt[3]{100n}$	$1/n$
61	3721	226 981	7.810 25	3.936 44	24.698 18	8.480 74	18.271 07	0.016 39
62	3844	238 328	7.874 01	3.957 84	24.899 80	8.526 84	18.370 37	0.016 13
63	3969	250 047	7.937 25	3.979 00	25.099 80	8.572 43	18.468 61	0.015 87
64	4096	262 144	8.000 00	4.000 00	25.298 22	8.617 55	18.565 81	0.015 63
65	4225	274 625	8.062 26	4.020 67	25.495 10	8.662 20	18.662 01	0.015 38
66	4356	287 496	8.124 04	4.041 18	25.690 47	8.706 40	18.757 22	0.015 15
67	4489	300 763	8.185 35	4.061 49	25.884 36	8.750 15	18.851 48	0.014 93
68	4624	314 432	8.246 21	4.081 60	26.076 81	8.793 47	18.944 81	0.014 71
69	4761	328 509	8.306 62	4.101 51	26.267 85	8.836 36	19.037 22	0.014 49
70	4900	343 000	8.366 60	4.121 23	26.457 51	8.878 85	19.128 75	0.014 29
71	5041	357 911	8.426 15	4.140 76	26.645 83	8.920 93	19.219 41	0.014 08
72	5184	373 248	8.485 28	4.160 11	26.832 82	8.962 16	19.309 22	0.013 89
73	5329	389 017	8.544 00	4.179 28	27.018 51	9.003 92	19.398 20	0.013 70
74	5476	405 224	8.602 33	4.198 28	27.202 94	9.044 84	19.486 37	0.013 51
75	5625	421 875	8.660 25	4.217 10	27.386 13	9.085 40	19.573 76	0.013 33
76	5776	438 976	8.717 80	4.235 76	27.568 10	9.125 60	19.660 37	0.013 16
77	5929	456 533	8.774 96	4.254 26	27.748 87	9.165 45	19.746 22	0.012 99
78	6084	474 552	8.831 76	4.272 60	27.928 48	9.204 96	19.831 33	0.012 82
79	6241	493 039	8.888 19	4.290 78	28.106 94	9.244 13	19.915 72	0.012 66

Solution

(a) $\sqrt{70}$ can be obtained directly from column 4 of the table giving

$\sqrt{70} = 8.37$ to three significant figures

(b) $\sqrt{72} = 8.485\,28$

$\sqrt{73} = 8.544\,00$

As 72.5 is half-way between 72 and 73 then $\sqrt{72.5}$ will be approximately half-way between 8.485 28 and 8.544 00.

$\sqrt{72.5} \approx (8.485\,28 + 8.544\,00)/2 = 8.514\,64$

Hence $\sqrt{72.5} \approx 8.51$ to two decimal places. Check your answer using your calculator.

(c) We look at the final column of the table to find the value of n that gives the largest $1/n$ value smaller than 0.014. This is $n = 72$, giving 0.013 89.

Next we find the value of n that gives the smallest $1/n$ value larger than 0.014 which is $n = 71$, and $1/n = 0.014\,08$.

The number that needs to be added to 0.013 89 to make 0.014 is 0.000 11. The total difference between 0.013 89 and 0.014 08 is 0.000 19.

An estimate of n that gives $1/n = 0.014$ is

$72 - (0.000\,11/0.000\,19) = 72 - 11/19 = 71.4$

What answer does your calculator give?

Worked example

5.5   Table 5.10 comes from a holiday brochure distributed by a leading travel company in Spring 1997.

I'm sorry Mother, he says we can't afford to take you.

Table 5.10. Madeira holidays

FLIGHTS AVAILABLE FROM 6 UK AIRPORTS (SEE PAGES 344–347)	**£10 UK departure tax per person included**									
Name & Board	**RAGA** Bed & Breakfast			**MADEIRA** Bed & Breakfast			**MONTE VERDE** Bed & Breakfast			
Flights	Mon Flights			Mon Flights			Mon Flights			
Code	OFR			OFD			OFW			
Prices based on	PB WC BL			PB WC			PB WC			
Nights	7	14	All	7	14	All	7	14	All	
Adult/Child	Adult	Adult	1st Ch	Adult	Adult	1st Ch	Adult	Adult	1st Ch	
01 Nov - 07 Nov	299	425	179	285	409	179	285	405	179	
08 Nov - 13 Nov	289	415	169	275	399	159	275	395	159	
14 Nov - 20 Nov	265	385	149	249	369	149	249	365	149	
21 Nov - 28 Nov	249	365	149	235	345	149	235	345	149	
29 Nov - 12 Dec	255	355	139	239	339	139	239	335	139	
13 Dec - 17 Dec	235	576	139	219	559	139	219	555	139	
18 Dec - 24 Dec	519	695	419	505	679	419	505	675	419	
25 Dec - 31 Dec	449	495	289	435	479	289	435	475	289	
01 Jan - 05 Feb	275	405	139	259	389	139	259	385	139	
06 Feb - 12 Feb	285	425	189	269	409	189	269	405	189	
13 Feb - 25 Feb	305	435	159	289	419	159	289	415	159	
26 Feb - 27 Mar	329	455	189	315	439	189	315	435	189	
28 Mar - 03 Apr	345	485	229	329	465	229	329	465	229	
04 Apr - 08 Apr	395	525	299	379	505	299	379	505	299	
09 Apr - 23 Apr	349	475	249	335	459	249	335	455	249	
Supplements per person per night	Single Room £11.40 Sea View £2.96 Full Board £21.13 Half Board £11.95			Balcony £1.20			Balcony £1.25 Balcony and Sea View £1.75			
Reductions per person per night	One Adult only sharing £4.30									

Earlybird Savings
Save UP TO **£80**
Per couple on listed prices
Savings apply to a wide range of holidays throughout the season
See page 2 for details
Ask your **TRAVEL AGENT** for today's savings

2nd child pays 1st child price plus £29 per week

Source: Thomson Holidays Limited

(a) What will be the cost of seven nights in the Hotel Monte Verde for a couple travelling on Monday, 1 December if they want a balcony and sea view but do not qualify for any Earlybird Savings?

(b) How much extra would the same holiday cost if they took their two children?

Solution (a) Basic holiday cost is 2(£239) = £478
Supplement for balcony and sea view = 2(7)(£1.75) = £24.50
Hence total cost is £478.00 + £24.50 = £502.50

(b) Extra cost for two children is £139.00 + £139.00 + £29.00 = £307.00

Self-assessment questions 5.3

1. For the data in Table 5.7 what would be the total cost of hiring a Transit van (code 12 BV) for three days with CDI if the van is to be delivered to you and you live 8 miles from the nearest UK office?

2. For the data in Table 5.8 calculate the Remarriages as a percentage of all marriages, accurate to three decimal places, for 1991 (Table value is 36%).

Exercise 5.3

1. Use Table 5.10 to price a holiday for a married couple taking the wife's mother for a two week holiday starting on Monday 17 November. They all need full board and the mother needs a single room with a sea view (ignore the Earlybird Savings).

2. Use Table 5.9 to find $\sqrt[3]{6700}$ and compare this result with the result from your calculator.

3. Table 5.11 shows the monthly repayments to pay off various secured loans over certain time periods when the interest rate is 12.9% (APR).

Table 5.11. Monthly loan repayments

☐ With Personal Loan Repayments Insurance   ☐ Without Personal Loan Repayments Insurance

Amount of loan £	Secured									
	Repayment term									
	25 years		20 years		15 years		10 years		5 years	
	With	Without	With	Without	With	Without	With	Without	With	Without
3 000	35.40	32.13	36.95	33.54	40.18	36.47	47.88	43.46	73.93	67.10
5 000	59.00	53.55	61.58	55.89	66.97	60.78	79.80	72.43	123.21	111.83
7 000	82.60	74.97	86.22	78.25	93.76	85.10	111.72	101.40	172.50	156.56
10 000	118.00	107.10	123.17	111.79	133.95	121.57	159.61	144.86	246.43	223.66
15 000	177.00	160.65	184.75	167.68	200.91	182.35	239.41	217.29	369.64	335.49
20 000	236.01	214.20	246.33	223.57	267.88	243.13	319.21	289.72	492.86	447.32
25 000	295.01	267.75	307.92	279.47	334.85	303.91	399.02	362.15	616.07	559.15

(a) Use this table to find the monthly repayments for a loan of £10 000 over 15 years without Personal Loan Repayments Insurance.

(b) How much more would this be each month if you had Personal Loan Repayments Insurance?

(c) How much in total would be paid over the 15 years for the loan in part (a)?

(d) Determine the total cost of the Personal Loan Repayments Insurance over the 15 years for the loan in part (a).

Test and assignment exercises 5

1. A survey took place to investigate the prices of oranges in a certain town. The investigator went into 15 greengrocer shops on one Saturday and the prices she observed are recorded in Table 5.12. Use these data to make a frequency table for the price of oranges.

Table 5.12. Orange prices (pence)

Shop	A	B	C	D	E	F	G	H	I	J	K	L	M	N	O
	17	15	18	20	15	15	11	15	19	18	15	17	12	13	12

2. Table 5.13 shows the average attendance at football and rugby league matches from 1971/2 to 1993/4.

Table 5.13. Average attendance at football and rugby league matches (GB)

	Football Association[1] Premier League[2]	Football League[1] Division One[3]	Scottish Football League Premier Division[4]	Rugby Football League Premier Division
1971/72	31 352	14 652	5 228	–
1976/77	29 540	13 529	11 844	–
1981/82	22 556	10 282	9 467	–
1986/87	19 800	9 000	11 720	4 844
1989/90	20 800	12 500	15 576	6 450
1990/91	22 681	11 457	14 424	6 420
1991/92	21 622	10 525	11 970	6 511
1992/93	21 125	10 641	11 520	6 170
1993/94	23 040	11 777	12 351	5 683

[1] League matches only until 1985/86. From 1986/87, Football League attendances include promotion and relegation play-off matches.
[2] Prior to 1992/93, Football League Division One.
[3] Prior to 1992/93, Football League Division Two.
[4] Prior to 1976/77, Scottish League Division One.

Sources: Football Association Premier League; Football League; Scottish Football League; Rugby Football League.

(a) What was the average attendance at League Division One matches in 1992/93?
(b) In which year and for which type of match was the average attendance at football matches the highest?
(c) In which year and for which type of match was the average attendance at football matches the lowest?

3. Table 5.14 shows the number (in thousands) of births in the United Kingdom.

Table 5.14. Live births in UK (thousands)

	Live births				Rates				
	Total	Male	Female	Sex ratio	Crude birth rate[1]	General fertility rate[2]	TPFR[3]	Still-births[4]	Still-birth rate[4]
1900–02	1 095	558	537	1 037	28.6	115.1	–	–	–
1910–12	1 037	528	508	1 039	24.6	99.4	–	–	–
1920–22	1 018	522	496	1 052	23.1	93.0	–	–	–
1930–32	750	383	367	1 046	16.3	66.5	–	–	–
1940–42	723	372	351	1 062	15.0	–	1.89	26	–
1950–52	803	413	390	1 061	16.0	73.7	2.21	18	–
1960–62	946	487	459	1 063	17.9	90.3	2.80	18	–
1970–72	880	453	427	1 064	15.8	82.5	2.36	12	13
1980–82	735	377	358	1 053	13.0	62.5	1.83	5	7
1979	735	378	356	1 061	13.1	64.1	1.86	6	8
1980	754	386	368	1 050	13.4	64.9	1.89	6	7
1981	731	375	356	1 053	13.0	62.1	1.81	5	7
1982	719	369	350	1 054	12.8	60.6	1.78	5	6
1983	721	371	351	1 058	12.8	60.2	1.77	4	6
1984	730	373	356	1 049	12.9	60.3	1.77	4	6
1985	751	385	366	1 053	13.3	61.4	1.80	4	6
1986	755	387	368	1 053	13.3	61.1	1.78	4	5
1987	776	398	378	1 053	13.6	62.3	1.82	4	5
1988	788	403	384	1 049	13.8	63.2	1.84	4	5
1989	777	398	379	1 051	13.6	62.4	1.81	4	5
1990	799	409	390	1 049	13.9	64.2	1.84	4	5
1991	793	406	386	1 052	13.7	63.6	1.82	4	5
1992	781	400	380	1 052	13.5	63.4	1.80	3	4
1993[5]	762	391	371	1 054	13.1	62.4	1.76	4	6

[1] Rate per 1000 population.
[2] Rate per 1000 women aged 15–44.
[3] Total fertility rate is the average number of children which would be born per woman if women experienced the age-specific fertility rates of the period in question throughout their child-bearing life span. UK figures for the years 1970–72 and earlier are estimates.
[4] Figures given are based on stillbirths of 28 completed weeks gestation or more. On 1 October 1992 the legal definition of a stillbirth was altered to include babies born dead between 24 and 27 completed weeks gestation. Between 1 October and 31 December 1992 there were 258 babies born dead between 24 and 27 completed weeks gestation. If these babies were included in the stillbirth figures given, the stillbirth rate would be 5.
[5] Later figures not available.

Source: Annual Abstract of Statistics 1995. Office for National Statistics. Crown Copyright, 1997.

(a) Determine the total number of births in the United Kingdom over the years 1980 to 1984 inclusive.

(b) How many more male births than female births were there in 1990?

(c) Write two comments/observations from these data.

4. Using any page of text such as page 73, make a frequency table for the letters of the alphabet. Before starting can you guess the commonest letter and the rarest letter? Were you right?

6 Charts

Objectives	After reading this chapter, you should be able:
	• to interpret the information shown in common charts
	• to choose a suitable chart type to display data
	• to produce simple charts

6.1 Introduction

The main purpose of a chart or diagram is to convey information simply, clearly, quickly and in such a way that merely presenting the raw data in a table does not do. There are a few characteristics that all charts or displays of data must have in order for them to be properly interpreted. Above all else they should be easy to understand and not too complicated. Most common charts are either time series charts or charts showing the frequency of occurrence of different factors.

KEY POINT

All charts should have the following characteristics:
- There must be a title.
- All units of measurement must be clearly stated.
- There must be a key to any symbols or shading used.
- The source of the data must be given if it is not obvious.
- Any ambiguous or non-standard terms used must be defined, usually in a footnote.
- The diagram must be sufficiently large for the detail to be clear.

6.2 Time series charts

Whenever one of the variables on a chart is time, such as years, seconds or months, it is usual to plot this time variable on the horizontal axis. Consider the chart shown in Figure 6.1, which clearly shows that there is

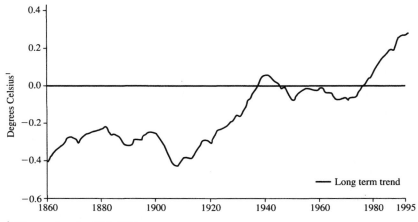

Figure 6.1.
Global temperature
variations.
Source: Department of
the Environment.

¹Difference from the 1961–1990 average

some evidence for global warming, although the apparent increase in temperature is less than one degree Celsius. This chart satisfies all the requirements of a good chart, and conveys the underlying pattern contained in the raw data much better and quicker than the data alone could have done.

Figure 6.2 is another example of a clear diagram which shows that for the past several decades there have been more births than deaths. The projections also show that the government statisticians expect there to be more deaths than births after about 2025.

Both of the previous charts showed how a particular quantity varied over time, and as such are both examples of time series, and the convention when displaying any time series is to have time on the horizontal axis as in Figures 6.1 and 6.2.

KEY POINT

- Time series charts usually have time on the horizontal axis.
- Leave enough space to the right of the time axis to allow for the future to be estimated if this is relevant.

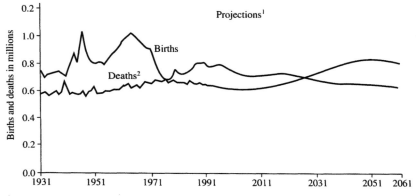

Figure 6.2.
Births and deaths (UK).
Sources: Office of
Population Censuses and
Surveys; Government
Actuary's Department;
General Register Office
(Scotland); General
Register Office (Northern
Ireland). Crown
Copyright, 1997.

[1] 1992-based projections
[2] Includes deaths of non-civilians and merchant seamen who died outside the country

---

Self-assessment question 6.2

1. If you were to draw a chart to show how the total monthly sales of a company have changed over the past three years, how would you label the horizontal axis?

Exercise 6.2

1. When in the 20th century was the mean global temperature the lowest? (Use Figure 6.1.)

2. When in the past 50 years were the birth and death rates the closest? (Use Figure 6.2.)

3. The following data are the monthly sales figures (thousands sold) for one brand of tinned meat, starting with the sales for January 1995. Draw a suitable time series chart and comment.

31 33 29 23 17 19 16 20 17 25 33 36
38 40 31 23 24 19 22 26 34 38 44 48

6.3   Frequency charts

One of the commonest types of chart is called a *histogram*. Figure 6.3 is a histogram showing how the heights of men vary. As with all histograms the area of each block represents the frequency for the interval on which it stands, and if all the intervals are the same width, as they are here, then the height of each block can be taken to represent the frequency. It can be seen from Figure 6.3 that few men are smaller than 160 cm (5ft 3in) or taller than 185 cm (6ft 1in) and the most 'popular' or common height is just less than 175 cm (5ft 9in).

Figure 6.4 is a good example of a *pie-chart,* where a circle or disc is divided into segments whose size, as measured by the area (and hence the angle at the centre) represents the proportion or percentage associated with the label of the segment. It can be seen from this pie-chart that more than twice as many children's videos are bought than

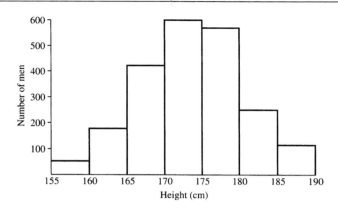

Figure 6.3.
Height distribution of
men aged 16–24. *Source:*
Adult Heights and
Weights Survey, 1980.
Office for National
Statistics. Crown
Copyright, 1997.

Figure 6.4.
Pre-recorded video
sales: by type, 1995

videos based on TV, while feature films are the most popular. (What do
you think would come under the heading 'Other'?)

In a pie-chart it is often difficult to judge which of two segments is
the smallest, and so the actual figures or percentages should always be
given to allow for such comparisons. Here the diagram does not make it
clear that the sales of music videos is slightly higher than the sales of
other videos, but the given figures make this clear.

When drawing this pie-chart, the angle of the segment for television (12%) had to be evaluated by calculating 12% of the 360° in a full circle. Following the method of Chapter 3, this angle is

$$\frac{12}{100} \times 360 = 43.2°$$

Another common chart is a bar chart and Figure 6.5 is an example. This chart is an example of a *multiple bar chart*, similar to a histogram in that height represents frequency or popularity but where the variable along the horizontal axis is not one measured on a continuous scale such as height or age.

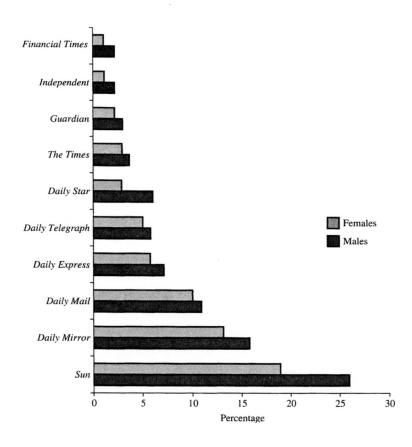

Figure 6.5.
Reading of national newspapers by sex, 1995

Worked example

6.1 Consider the information provided in Figure 6.5.

(a) Of the newspapers mentioned, which is read by the fewest people?
(b) Which newspaper is read by the most people?
(c) Which newspapers could be called 'a male' paper?

Solution   To find the percentage of people who read any particular newspaper, the two percentages for each of the sexes need to be added. However, despite this complication, the following answers are clear from the diagram:

(a) The *Financial Times* and the *Independent* have the fewest readers, with the *Guardian* and *The Times* also having few readers.
(b) The *Sun* has the most readers.
(c) The *Daily Star* has twice as many male readers as female readers, and the *Sun* could also be called 'male'.

Note that the percentages add up to more than 100%. This means that some people read more than one newspaper.

KEY POINT

> When the frequencies to be charted relate to data that can take many different values, such as heights or costs, draw a **histogram**.
>
> When the frequencies to be charted relate to data that can take only a few distinct values which can be ordered or which are of different categories, draw a **bar chart**.
>
> When the frequencies to be charted relate to data that can take only a few distinct values with no obvious order, draw a **pie chart**.

There are many other types of charts and the following provides examples of some of them.

Worked example

6.2   Figure 6.6 shows the age structure of the United Kingdom in 1995. It is not a common type of chart, but it allows for many interesting observations. It clearly shows that for either sex the number of people alive in 1995 who were born before 1915, i.e. aged over 80 in 1995, is very low, as expected. Study this chart carefully and then make at least three other comments on what this chart indicates.

Solution   (a) Females tend to live longer than males; there are many more old women than old men.
(b) There is a sharp increase in the number of old people in their late 70s or an increase in the birth rate in 1920, probably caused by men returning home after the First World War.
(c) There was a similar increase in the birth rate after the Second World War, i.e. people aged about 48–50.

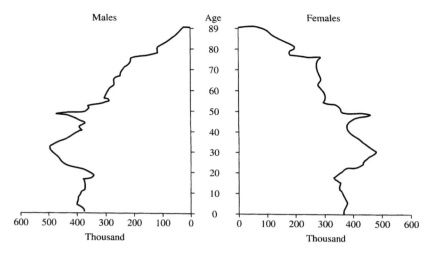

Figure 6.6.
Population: by age and
sex 1995 (United
Kingdom). *Source*: *Social
Trends* 1996, Office for
National Statistics;
General Register Office
(Scotland); General
Register Office (Northern
Ireland). Crown
Copyright, 1997.

(d) There was a decrease in the birth rate during 1916–19, caused by the
absence of men away from home, fighting in the First World War.

(e) The fall in the birth rate during the Second World War was not as
great as it had been in the First World War. This may have been due
to better transportation which allowed more frequent leave, or
because there were more men in the armed forces stationed in Britain!

(f) There was a 'bulge' in the birth rate around 1965, which has
produced a 'bulge' in the number of people of both sexes who were
30 years old in 1995.

Self-assessment question 6.3

1. What type of chart would you draw for each of the following sets of data?
    (a) A sample of 200 lifetimes for tested light bulbs
    (b) The religion of a sample of 100 adults attending a sports centre.
    (c) The number of siblings (brothers and sisters) that a sample of 50 children had. (If you
        are an only child then you have no siblings.)

Exercise 6.3

1. Use Figure 6.3 to estimate the proportion
    of men aged between 16 and 24 who are
    over 6 feet tall (6 feet is about 183 cm).

2. If the data that were used to produce
    Figure 6.4 were for sales of 4000 videos,
    how many music videos were there?

3. What percentage of the population reads
    the *Sun*? (Use Figure 6.5.)

4. Use the data in Table 5.8 to draw a
    multiple bar chart comparing the
    total number of marriages where it
    was the first marriage for one partner
    only with those marriages where it
    was the second or subsequent marriage
    for both partners. Put years on the
    horizontal axis and have two bars for
    each year.

Test and assignment exercises 6

1. Draw a histogram for the following data on house prices taken from question 2 in Exercise 5.2.

Class £000s	Frequency
0 to under 40	7
40 to under 80	14
80 to under 120	6
120 to under 160	2
160 to under 200	0
200 or more	1

2. Use your local newspaper to collect the prices of 20 houses that are for sale and hence draw a simple histogram showing the price variation. Compare this with the results from question 1 above.

3. Study Figure 6.7, and then answer the following questions:

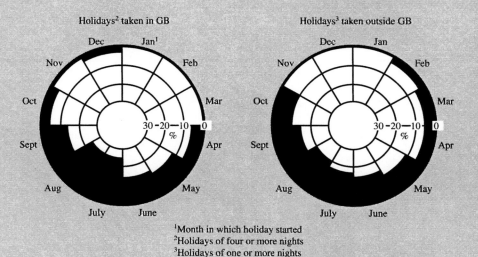

¹Month in which holiday started
²Holidays of four or more nights
³Holidays of one or more nights

Figure 6.7. Monthly distribution of holidays taken by adults resident in Great Britain, 1980. *Source: Social Trends*, No. 12, 1979, Office for National Statistics. Crown Copyright, 1997.

(a) For holidays taken in Great Britain, which months are the most popular, and which the least popular?
(b) For holidays taken outside Great Britain which month is the most popular and which months the least popular?
(c) Is it true to say more people take August holidays in Great Britain than outside?
(d) Table 6.1 shows the holiday pattern for 1995. How would you draw a chart to illustrate these data. Can you comment on whether holiday patterns have changed?

Table 6.1. Holidays taken by UK residents; by month 1995

	Short holidays[1]		Long holidays[2]		All
	Home (%)	Abroad (%)	Home (%)	Abroad (%)	Holidays (%)
January	5	6	1	4	3
February	7	5	2	4	4
March	7	12	3	5	5
April	10	11	7	7	8
May	11	10	10	11	11
June	9	6	9	12	10
July	9	8	17	14	13
August	10	8	23	16	16
September	9	9	10	10	10
October	9	10	7	8	11
November	5	11	2	4	3
December	8	5	6	5	4
All months	100	100	100	100	· 100

[1] Holidays of one to three nights.
[2] Holidays of four nights or more.

Source: British Tourism Authority.

4. Use the data from Table 6.2 to draw a suitable chart comparing the attendance at football matches for the Premier League and Division One since 1989/90.

Table 6.2. Average attendance at football and rugby league matches (Great Britain)

	Football Association[1] Premier League[2]	Football League[1] Division One[3]	Scottish Football League Premier Division[4]	Rugby Football League Premier Division
1971/72	31 352	14 652	5 228	–
1976/77	29 540	13 529	11 844	–
1981/82	22 556	10 282	9 467	–
1986/87	19 800	9 000	11 720	4 844
1989/90	20 800	12 500	15 576	6 450
1990/91	22 681	11 457	14 424	6 420
1991/92	21 622	10 525	11 970	6 511
1992/93	21 125	10 641	11 520	6 170
1993/94	23 040	11 777	12 351	5 683

[1] League matches only until 1985/86. From 1986/87, Football League attendances include promotion and relegation play-off matches.
[2] Prior to 1992/93, Football League Division One.
[3] Prior to 1992/93, Football League Division Two.
[4] Prior to 1976/77, Scottish League Division One.

Sources: Football Association Premier League; Football League; Scottish Football League; Rugby Football League.

5. Study Figure 6.8, and then answer the following questions:
   (a) Which age group spend the least time in paid work?
   (b) Which age group spend the most time in paid work?
   (c) Other than the 60 and over group, are there any other differences?

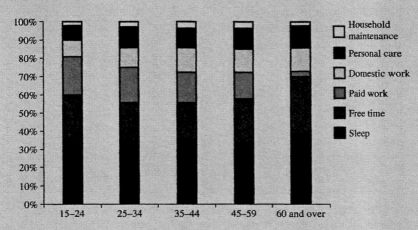

Figure 6.8. Time use by age, May 1995 (Great Britain)

6. Draw a suitable chart to illustrate the data in Table 6.3 and comment on this chart.

Table 6.3. Television viewing; by sex and age 1995 in the United Kingdom (hours and minutes per week)

	Males	Females
4–15	18 : 51	17 : 14
16–24	18 : 03	20 : 16
25–34	22 : 16	25 : 54
35–44	22 : 51	25 : 10
45–54	24 : 11	26 : 22
55–64	28 : 02	31 : 23
65 and over	34 : 29	36 : 42
All aged 4 and over	23 : 45	26 : 25
Reach[1] (percentages)		
Daily	81	83
Weekly	96	96

[1] Percentage of the population aged 4 and over who viewed TV for at least three consecutive minutes.

Source: Broadcasters' Audience Research Board; British Broadcasting Corporation; AGB Limited.

7. Study Figure 6.9.
   (a) Where on this chart will you be in five years' time?
   (b) Is there sexual equality on earnings?

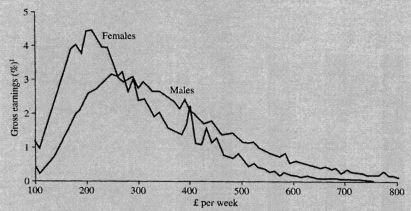

¹Full-time employees on adult rates whose pay was not affected for the survey period by absence

Figure 6.9. Average gross earnings; by sex, April 1995 (UK). *Source:* New Earnings Survey, Office for National Statistics and Department of Economic Development, Northern Ireland, 1996. Crown Copyright, 1997

7 Co-ordinates

Objectives

After reading this chapter, you should be able:

- to understand the common two-dimensional co-ordinate system
- to plot points using a co-ordinate system
- to extract information from charts that use a co-ordinate system

7.1 Introduction

Many methods of displaying data require an understanding of co-ordinates in both statistics and mathematics. This chapter is on the commonest co-ordinate system in two dimensions, sometimes referred to as the Cartesian co-ordinate system.

7.2 Co-ordinates

Co-ordinated?

Table 7.1 shows the scores in six premier-division football matches played one Saturday in December 1996. One convenient method of

Table 7.1.
Scores in six football
matches

Match number	(i)	(ii)	(iii)	(iv)	(v)	(vi)
Home team	1	1	1	3	2	0
Away team	0	2	1	1	0	1

illustrating the data is to use a co-ordinate system in which each match is represented by a point or a dot. The position of this point is determined by the number of goals scored by the two sides. A suitable co-ordinate system is shown in Figure 7.1 where the six matches or points have been placed in the appropriate position. For example, match number (iv) is represented by a point which has the co-ordinates: Home goals 3, Away goals 1. The point when both teams score 0 is called the *origin*.

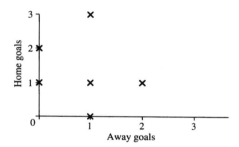

Figure 7.1.
Results of six football
matches

Do these six results show a trend? Not clearly, but perhaps more matches would show that the home team tends to score more goals than the away team. You might test this by plotting the results of some recent matches.

The data used in this example were simple (only the values 0, 1, 2, 3 were used), but a co-ordinate system together with suitable axes can be used to display data of any magnitude. The data shown in Table 7.2 were taken from seven adults, and were used as an attempt to investigate the relationship between height and head circumference measured around the head just above the eyes.

Table 7.2.
Height and head
circumferences of
seven adults

Person	A	B	C	D	E	F	G
Height (cm)	182	185	175	168	166	165	161
Head circumference (cm)	61	59	58	57	56	62	55

The two axes that should be used to display these data are height and head circumference, but since all heights are greater than 160 cm, there is no need to start the height axis at 0 cm since this would waste a lot of space. One possible method is to start the height axis at 150 cm and the head circumference axis at 50 cm, as is shown in Figure 7.2. These points show a clear pattern which is broken by person F, who either has short legs or a big head!

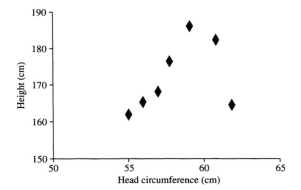

Figure 7.2.
Height and head
circumference for
seven adults

It is common in many situations like this to use symbols for the two variables involved. If we let x represent head circumference and y represent height, then the axes used are referred to as the y axis and x axis, and person A has co-ordinates $x = 61$, $y = 182$. This is often abbreviated by saying person A has co-ordinates (61,182), where the convention is to give the x co-ordinate first. It is also convention to have the x axis horizontal and the y axis vertical.

KEY POINT

> The usual way to represent a point using two co-ordinates is by using x and y with x first. So the point represented by $x = 2$, $y = 6$ is written (2,6).
>
> When plotting such points it is conventional to have x on the horizontal axis and y on the vertical axis.

Worked example

7.1   It is clearly reasonable to expect some relationship between head circumference and height. One suggested relationship is that height is three times the head circumference. Test this relationship using the head circumference of the seven people discussed earlier.

Solution The first step is to calculate the expected height if the relationship is true, i.e. multiply each head circumference by three. This produces the results in Table 7.3.

Table 7.3.
Head circumference and predicted height of seven adults

Person	A	B	C	D	E	F	G
Head circumference (cm)	61	59	58	57	56	62	55
Height predicted (cm)	183	177	174	171	168	186	165

These seven points are then plotted on suitable axes, together with the original points. See Figure 7.3. This clearly shows that the claim is reasonable, but again emphasizes that person F is unusual. Such points are called outliers.

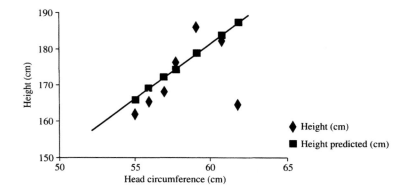

Figure 7.3.
Actual compared with predicted head circumferences

Worked example

7.2 This example of co-ordinates is taken from computer graphics. Most computer graphics languages use a conventional co-ordinate system with a horizontal x axis, a vertical y axis and the notation that the point (2,3) is at $x = 2$, $y = 3$. The point (0,0), usually called the origin, is situated at the bottom left-hand corner of the screen. The two commands used in one particular system are:

```
MOVE X,Y
```

which moves the imaginary pen or cursor to the point (X,Y) but no line or point is drawn, and

```
DRAW X,Y
```

which causes a line to be drawn from the current position of the cursor to the new position X,Y.

What shape would the following commands produce?

```
MOVE  10,10
DRAW  110,10
DRAW  110,60
DRAW  10,60
DRAW  10,10
MOVE  10,60
DRAW  60,80
DRAW  110,60
```

Solution First a co-ordinate system is needed with x and y axes, with x varying from 0 to 110 and y varying from 0 to 80. The results of following the above commands using graph paper are shown in Figure 7.4, but on the monitor screen only the house shape would be drawn; no axes would be seen.

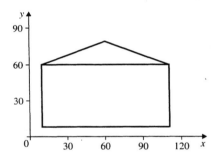

Figure 7.4.
Computer graphics
example

Worked example

7.3 This example is not applied to any particular practical problem, but is one that will often occur in more advanced areas of study.

Using a suitable co-ordinate system illustrate graphically the relationship between y and x if $y = 2x + 1$. Use values of x between -4 and $+4$.

Solution We will evaluate pairs of numbers (x,y) so that each pair satisfies this relationship. For example if $x = 3$, then $y = 2(3) + 1 = 6 + 1 = 7$. Each pair, such as $(3,7)$, will then represent a point in a co-ordinate system. The result of evaluating some of these pairs is:

x	-4	-3	-2	-1	0	1	2	3	4
y	-7	-5	-3	-1	1	3	5	7	9

(If you are not sure about negative numbers read Chapter 1.)

The result of plotting these points can be seen in Figure 7.5, which shows the nature of the relationship. In Figure 7.5 the lowest point has co-ordinate $(-4, -7)$ and the highest has co-ordinate $(4, 9)$.

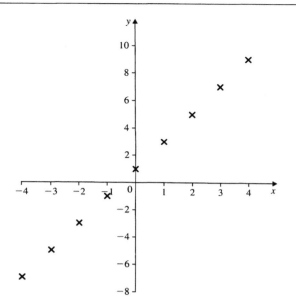

Figure 7.5.
The relationship
$y = 2x + 1$

Self-assessment questions 7.2

1. Measure your height and head circumference and plot your 'dot' on Figure 7.2.

2. Ten car drivers were each asked how many cars they had owned and how long they had held a full licence, with the following results:

Person	A	B	C	D	E	F	G	H	I	J
No. of cars	6	2	8	4	1	1	15	7	9	3
No. of years	10	1	15	23	1	2	4	10	17	4

Choose suitable axes, display the data, and briefly comment.

3. Using axes as in Figure 7.1 plot the results of five imaginary football matches in which the home team always won.

Exercise 7.2

1. It is claimed that there is a strong relationship in women between the size of their shoes and their hip measurements. Test this claim by measuring, recording and then plotting the data from a sample of six women.

2. What shape would be produced if the following instructions were obeyed by a computer with a graphics monitor?

MOVE   20,100   DRAW   80,100
MOVE   50, 30   DRAW   50,140
MOVE   45,140   DRAW   55,140

3. A book publisher has produced seven comparable books with the following costs:

Book	A	B	C	D	E	F	G
Quantity produced (x 000)	3	4.5	7	8	10	12	15
Manufacturing costs (£y 000)	14	19	30	33	40	50	60

Display the data graphically and comment.

Test and assignment exercises 7

1. Using x and y co-ordinates which both go from 0 to 150, write instructions for a computer so that a large capital letter F is drawn.

2. The six points A, B, C, D, E and F have (x, y) co-ordinates as follows:

   A: (1,3)   B: (4,2)   C: (2,1)
   D: (7,2)   E: (3,6)   F: (4,4)

   (a) Which point is nearest to the x axis?
   (b) Which point is nearest to the y axis?
   (c) Which point is furthest from (0,0), the origin?
   (d) Which point is nearest to point B? (You may need to use a ruler on your graph to decide.)

3. Using values of x from -5 to $+5$ inclusive, illustrate the relationship between y and x if $y = 4x - 2$.

4. Decide with reference to Figure 7.4, whether the following statements are true or false:
   (a) The point (40,30) is inside the rectangle.
   (b) The point (80,65) is inside the triangle.
   (c) The point (20,70) is inside the triangle.
   (d) The straight line from (80,20) to (0,80) cuts into the triangle.
   (e) The straight line from (10,10) to (110,60) divides the rectangle into two triangles of equal area.

5. Write instructions for a computer so that the word HEALTH is drawn. You can assume that the x and y co-ordinates both vary from 0 to 150, for each letter.

8 Graphs

Objectives

After reading this chapter, you should be able:

- to construct simple graphs showing the relationship between two variables
- to understand the meaning of dependent and independent variables
- to use graphs to produce estimates
- to estimate and interpret gradients from graphs

8.1 Introduction

Like the charts discussed in Chapter 6, one of the functions of a graph is to display data so that any pattern, trend or relationship can easily be seen. A graph, however, is a specific type of diagram that uses a co-ordinate system as described in Chapter 7 to show the relationship between two variables. The two variables may be related precisely, which is the case with an *equation* connecting the two variables (Chapter 13), or may have what is called an *empirical* relationship. An empirical relationship is one that is not always exactly true, but tends to be approximately true and is based upon many observations of the two variables. One example of an empirical relationship is that for normal adults their height is approximately three times the circumference of their heads measured around the forehead (see Chapter 7).

8.2 Graphs

Table 8.1.
Heights and weights of women aged 16 to 64, 1980

Is there an empirical relationship between a person's height and weight? The first step in investigating such a relationship is to collect heights and weights of a large sample of people. Table 8.1 shows some of the results of such a survey.

Height group (cm)	150.1–155.0	155.1–160.0	160.1–165.0	165.1–170.0	170.1–175.0
Group midpoint (cm)	152.55	157.55	162.55	167.55	172.55
Average weight (kg)	57.3	60.5	62.8	65.1	68.7

Source: Adult Height and Weight Survey, 1980. Office for National Statistics. Crown Copyright, 1997.

The basic relationship is clear from this table; that is, the taller a woman is, then, on average, the heavier she is, which is only what would have been expected. Any further comments and conclusions can best be made after a suitable graph has been drawn, showing the relationship between these two variables. Let the variable H stand for the midpoint of the height group, and W stand for the average weight of those women in any particular height group. Since the weight is the variable we are interested in (that is, to be predicted from height), it is a convention to choose a co-ordinate system with weight W on the vertical axis and height H on the horizontal axis. More formally, W is the *dependent* variable that depends upon the *independent* variable H.

KEY POINT

> The dependent variable is the variable that depends upon the independent variable. It is usual to plot the dependent variable on the vertical (y) axis and the independent variable on the horizontal (x) axis.

Having decided which way round to have the axes, the next problem is to decide upon a scale that will allow the graph to be drawn on the piece of paper available (preferably graph paper with 1 cm squares). The horizontal variable, or independent variable, H, varies from 152.55 cm to 172.55 cm, but to allow a little leeway at each end we can have H varying from 150 cm to 180 cm. To start H at zero would produce a tiny graph and a great deal of wasted paper! The weight axes can be started at 55 kg and go up to 70 kg, but since the data to be plotted represent the average weight, it would be better to let the weight vary from 40 kg to 90 kg so that any individual, overweight or underweight, can be plotted on the graph. Using these axes and the data in Table 8.1, the graph in Figure 8.1 can be drawn, where the points, when plotted, are joined together to show the relationship between height and average weight. The curve joining the points should be reasonably smooth, and pass through all the points where possible. If the data do not show such a smooth relationship then it is best to draw a smooth line which follows the general trend rather than try to join all the points.

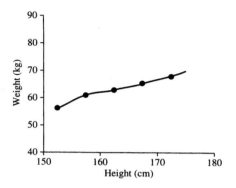

Figure 8.1.
Average weight (*W*) and height (*H*) of women aged 16–64, 1980.
Source: Adult Height and Weight Survey, 1980. Office for National Statistics. Crown Copyright, 1997.

KEY POINT

When empirical data are used to produce a graph it is not necessary to join all the points. Simply indicate the trend with a smooth line.

One advantage of this graph over the tabulated data is that the average weight for any particular height can be found. For example, women who are 165 cm tall (5ft 5in) weigh, on average, 63.5 kg (10 stone) and so a woman of 165 cm who weighs 70 kg (11 stone) could be said to be 6.5 kg (1 stone) overweight when compared with other women of her height.

Worked example

8.1 One of the commonest methods of buying a house is to have a repayment mortgage where the loan initially borrowed to buy the house is repaid over a fixed period of years by equal monthly repayments. Table 8.2 shows the monthly repayment needed to repay varying loans over a 25-year period when the net rate of interest is 8.925% p.a.

Table 8.2.
Calendar monthly repayments at 8.925% p.a.

Amount of loan (£)	1000	5000	10 000	20 000
Monthly repayments (£)	8.45	42.25	84.50	169.00

(a) Draw a graph to show the relationship between the size of loan and the monthly repayment. (The size of loan axis should go from zero to £40 000.)

(b) Use your graph to estimate the monthly repayments required for a loan of £16 500.

(c) Extend the graph to cover loans of up to £40 000 and hence estimate repayments for a loan of £37 950.

Solution   (a) See Figure 8.2 which shows a clear straight line or linear relationship.

(b) A loan of £16 500 will need a monthly repayment of £138 approximately. (A more accurate answer is £139.43 which can be found using the methods of Chapter 14, in which this example is discussed.)

(c) The straight line can easily be extended as shown in Figure 8.2 and this gives an estimate of £320.00. (Again the methods of Chapter 14 produce a more accurate answer of £320.68.)

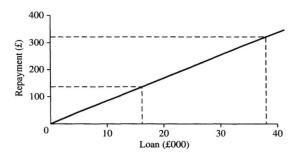

Figure 8.2.
Monthly repayments over 25 years at 8.925% p.a.

Worked example

8.2   A situation that faces many people is that of borrowing money to buy a car. How much can you afford to borrow? Table 8.3 is typical of many loan repayment tables and shows the various sums that can be borrowed over different repayment periods if the monthly repayment is fixed at £50.00. (The repayment amounts depend upon the interest rate applied, which vary considerably.)

Table 8.3.
Loan and repayment term for a monthly repayment of £50.00

Repayment term (months)	Loan (£)
12	540.54
18	772.56
24	983.48
30	1176.47
36	1356.55
48	1621.53
60	1875.11

(a) Graph the data and comment on the relationship.

(b) Use the graph to estimate the loan possible for a £50.00 monthly repayment over 42 months.

Solution   (a) See Figure 8.3. This is not a straight line, although it does demonstrate a clear, precise relationship. To see that it is not a straight line hold the graph to your eye and look along it, so exaggerating the curve. The gradient is decreasing as the term increases.

(b) When the term is 42 months the initial loan is approximately £1480.00.

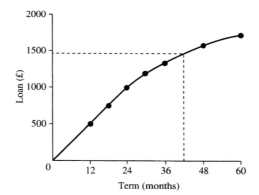

Figure 8.3.
Loan and repayment term
for a monthly repayment
of £50.00

KEY POINT   When the points to be plotted to produce the graph have been obtained from a formula then the graph should pass through every point. This can often show up an incorrectly calculated point.

Self-assessment questions 8.2

1. Use Figure 8.1 to estimate the average weight for women who are 180 cm tall.
2. Use Figure 8.3 to estimate the repayment term for a loan of £1000.00 with a monthly repayment of £50.00.

Exercise 8.2

1. Table 8.4 gives, for each height, two weights between which 80% of women of that height were found to weigh.

Table 8.4. Heights and weights of women aged 16–64, 1980

Height group (cm)	150.1–155.0	155.1–160.0	160.1–165.0	165.1–170.0	170.1–175.0
Group midpoint (cm)	152.55	157.55	162.55	167.55	172.55
80% are between	45.0	48.5	52.0	52.5	57.0
and (kg)	70.5	73.0	76.0	78.0	81.0

Source: Adult Height and Weight Survey, 1980. Office for National Statistics. Crown Copyright, 1997.

(a) Draw the graph in Figure 8.1 and add to it the information in Table 8.5.
(b) Record the height and weight of three women that you know, mark them on your graph and comment. (Perhaps not to them!)

2. Table 8.5 shows the number of UK males and females out of 100 000 surviving to a given age.

Table 8.5. UK life table

Age	Males	Females
0	100 000	100 000
5	96 186	97 019
10	95 866	96 794
15	95 601	96 608
20	95 151	96 300
30	93 820	95 311
40	91 968	93 778
50	87 591	90 656
60	75 823	83 646
70	52 350	67 835
80	21 130	36 118
90	2184	6079
100	23	161

(a) Draw a graph showing the survival patterns of the two sexes.
(b) Comment on the main features of the graph.
(c) Predict the number of males and females surviving to the age of 45.

8.3   Gradients

The *gradient* of a curve is the rate of change of the variable on the vertical axis with respect to the variable on the horizontal axis. If a graph is drawn with distance travelled on the vertical axis and time taken to travel this distance on the horizontal axis, then the gradient at any point is the rate of change of distance with respect to time; that is, speed. Consider Figure 8.4 which shows the progress of a canal barge between two locks 300 m apart. The gradient at the start and end of the journey is zero (flat), showing that the barge started from rest and stopped when it reached the next lock. For the first minute it picked up speed and then maintained this steady speed for a further two minutes. The next two minutes were spent at a slower speed before slowing down and finally stopping after a total journey of 300 m.

The gradient or slope of a curve at any particular point can be estimated from a graph by first drawing a tangent to the curve at that point, i.e. a straight line that just touches the curve but does not cut it, and then measuring the gradient of this tangent. The gradient of a straight line is the ratio of the increase along the vertical axis to the corresponding increase along the horizontal axis.

To estimate the gradient or speed of the barge after two minutes, draw a tangent to the curve at two minutes, as in Figure 8.4. This line has gradient

$$CB/AB = 265/2 = 132.5 \, \text{m/min}$$

Figure 8.4.
Distance against time for
a canal barge

This is equivalent to just less than 8 km/h, or a very fast walk. In the
examples that follow, gradients are further investigated with examples
of zero gradients (flat) and negative gradients (curves going down
instead of up).

KEY POINT	The gradient of a line drawn using the usual (x,y) co-ordinates is defined as the ratio of the change in y to the change in x. If the line is going up from left to right then the gradient is positive.

Worked example

8.3 Before mass-produced products such as transistors or light bulbs are
allowed to leave the factory it is usual to have some form of inspection.
If a batch fails the inspection, then all faulty or defective items are
replaced by sound ones, and so the whole batch will contain no
defectives. If a batch passes the inspection, it may still contain
defectives since the inspection often involves testing only a small
sample. Table 8.6 shows the percentage of defectives leaving the factory,
after one type of inspection, and possible replacement, for various
values of the percentage defective produced in manufacture.
(a) Graph the data and comment.
(b) What is the worst percentage defective that leaves the factory?
(c) Estimate and interpret the gradient when the percentage defective is
  (i) 3%   (ii) 8%   (iii) 15%

	% Defectives manufactured	% Defectives outgoing
Table 8.6. Percentage defectives manufactured and outgoing after inspection	0	0.0
	2	1.8
	4	3.2
	6	3.9
	8	4.1
	10	4.0
	12	3.7
	14	3.2
	16	2.7
	18	2.2
	20	1.8
	25	1.0
	30	0.5
	40	0.1

Solution

(a) Figure 8.5 shows a smooth curve which gets nearer and nearer to the horizontal axis. If no defectives are made, then obviously no defectives leave. If many of the items produced are defective then most inspections will fail, leading to all defectives being replaced, that is, there will again be only a few defectives leaving.

(b) The curve clearly reaches a maximum of approximately 4.1%.

(c) (i) When 3% defective, gradient is $CB/AB = 3.0/4.5 = +0.67$. This is a positive gradient meaning that as the % defective manufactured increases so does the % defective outgoing increase.

(ii) When 8% defective, gradient is 0. The % defective outgoing is neither increasing nor decreasing.

(iii) When 15% defective, gradient is $-DE/EF = -3.2/12 = -0.27$. This last gradient is negative because as the % defective manufactured increases, the outgoing defectives decrease.

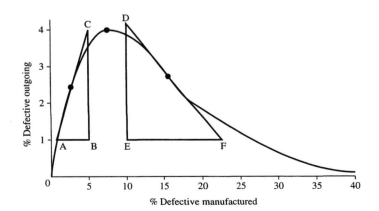

Figure 8.5. Percentage defective manufactured and outgoing after inspection

Self-assessment question 8.3

1. Use the graph in Figure 8.4 to estimate the speed of the barge after four minutes.

Exercise 8.3

1. Use the graph in Figure 8.4 to decide when the barge first started to slow down.

2. For which time is the gradient the highest in Figure 8.4?

3. Table 8.7 shows the temperature of a metallic body against time cooling in surroundings at 24 °C. Draw a graph showing the fall of temperature over the 15 minute period.

Table 8.7. Temperature of a metallic body cooling over time

Time (min)	0	1	2	3	4	5	6	7	8	9	10	11	12	13	14	15
Temperature (°C)	131.0	123.0	115.5	108.5	102.0	96.0	90.5	85.5	81.0	76.0	71.0	68.5	65.0	62.0	59.0	56.0

(a) Comment on the fall of temperature.
(b) Predict the temperature after 20 minutes.
(c) Determine the gradient after 2.5 minutes and 12.5 minutes.

Test and assignment exercises 8

1. Table 8.8 shows the cost per item for various batch sizes of a mass-produced item. Clearly the more that are made in one batch, the cheaper each one is to produce, since any initial set-up costs are shared among more items.
   (a) Draw the graph and comment.
   (b) Estimate the cost per item if a batch of 750 is made.
   (c) Estimate the cost per item if a batch of 5000 is made.
   (d) Estimate the gradients for batch sizes of 200, 500 and 1000.

Table 8.8. Costs for various batch sizes

Batch size	Cost per item
100	8.00
200	5.50
300	4.67
400	4.25
500	4.00
1000	3.50
2000	3.25

2. Table 8.9 shows various combinations of the amount advanced by a building society as a standard repayment mortgage and the term of years of the loan for a monthly repayment of £100.00.

Table 8.9. Loan and repayment term for a monthly repayment of £100.00

Repayment term (years)	Loan (£)
10	7 716.05
15	9 699.32
20	10 989.01
25	11 834.32
30	12 376.24

(a) Draw the graph and comment.
(b) With a monthly repayment of £100 over 18 years estimate the amount advanced.

3. The depth of water over a harbour bar at various times of day is:

Time (h)	07.00	08.00	09.00	10.00	11.00	12.00	13.00	14.00	15.00	16.00	17.00
Depth (m)	11.39	12.88	13.10	12.01	9.89	7.31	4.96	3.48	3.25	4.35	6.61

(a) Draw the graph of the data and comment.
(b) Estimate the times of high tide and low tide.
(c) During which times is the tide rising (positive gradient) and falling (negative gradient)?
(d) Estimate the greatest rate of decrease in depth and say when this occurs.

4. Complete the following table for the relationship between x and y when $y = 2 + 3x^2$ and hence draw a graph to illustrate this relationship. You should have x as the horizontal axis and y as the vertical axis.

x	0	1	2	3	4	5	6	7	8	9	10
y	2	5	14								

5. Draw a graph to illustrate the relationship between x and y if $y = 5 - 2x^2$. Use values of x from -5 to $+5$.

Section C
Statistics

9 Averages

Objectives	After reading this chapter, you should be able:
	• to explain the meaning and uses of the mean, median and mode
	• to calculate the mean for both raw data values and grouped data
	• to determine the median and mode for raw data values

9.1 Introduction

In daily life it is not unusual to meet phrases similar to

> The average cooked meal contains 800 calories

> The average lifetime of a light bulb is 850 hours

> The average number of children per family is 2.4

What does an average family size of 2.4 mean?

But is it clear what is meant by this term 'average'? Broadly speaking, the term is used to replace a collection of measurements by just one value. Indeed, the average of a set of numbers may be thought of as a number somewhere at the centre of the set and typical of the set as a whole. However, the word 'average' can be deliberately used to mislead or confuse as the word does have more than one meaning. In this chapter three measures of average are described; they are called the mean, the median and the mode.

9.2   Mean

Firstly, the mean, or the arithmetic mean as it is sometimes called, is found by adding the set of numbers together and dividing by the number of items involved. For example, the mean of the numbers 9, 5, 3, 5, 6 and 8 is

$$\frac{9+5+3+5+6+8}{6} = \frac{36}{6} = 6$$

Suppose the numbers are 50, 90, 60, 70, 60. These have a mean equal to

$$\frac{50+90+60+70+60}{5} = \frac{330}{5} = 66$$

To get a feel for this, the above can be interpreted in terms of the following: Melanie has 50p, Eleanor has 90p, Claire has 60p, Stephen has 70p and Ailsa has 60p. How much would each receive if they put their money into a pool and then took out an equal share of the pool? The amount of the pool would be $50 + 90 + 60 + 70 + 60 = 330$ pence. There are five of them to share it and so the share each would have is $330/5 = 66$ pence.

It is usual to represent the mean and its calculation by a mathematical shorthand.

KEY POINT

> In the calculation of the mean there are two operations:
>
> 1. Add all the numbers together.
> 2. Divide by the number of values.

In the above example there are five numbers 50, 90, 60, 70, 60. Suppose, more generally, there are n numbers, represented by

$$x_1, x_2, x_3, \ldots, x_n$$

The shorthand notation for 'add all the x-values together' is $\sum x$ (read 'sigma x' or 'sum of x' – $\sum$ is a capital S in Greek).

For example,

$$x_1 = 5, \quad x_2 = 12, \quad x_3 = 8, \quad x_4 = 5, \quad x_5 = 7, \quad x_6 = 4, \quad x_7 = 10$$
$$\sum x = 51$$

obtained using the key sequence

| 5 | + | 12 | + | 8 | + | 5 | + | 7 | + | 4 | + | 10 | = |

The sum is then divided by the number of items in order to find the mean of x. This is often written $\bar{x}$ (read 'x-bar').

> The mathematical shorthand for the mean of a set of data is
>
> $$\bar{x} = \frac{\sum x}{n}$$

If you come across the word 'average' in any technical publication or announcement it will probably be the mean that is being referred to.

Worked example

9.1 A survey took place to investigate the prices of apples and oranges in a certain town. The investigator went into 15 greengrocer shops on one Saturday and the prices he observed are recorded in Table 9.1.

Table 9.1.
Shop prices
(in pence)

	Shop														
	A	B	C	D	E	F	G	H	I	J	K	L	M	N	O
Apples (1 kg)	75	80	86	90	74	78	69	82	79	94	79	85	94	70	80
Oranges (each)	17	15	18	20	15	15	11	15	19	18	15	17	12	13	12

Determine

(a) the mean price of 1 kg of apples,
(b) the mean price per orange,

in these 15 shops.

Solution (a) Using the formula $\bar{x} = \dfrac{\sum x}{n}$ for the apple prices,

$$\sum x = 75 + 80 + 86 + \ldots + 80 = 1215$$

and so

$$\bar{x} = 1215/15 = 81$$

The mean price of 1 kg of apples in this town is 81p.

(b) For orange prices,

$$\sum x = 17 + 15 + 18 + \ldots + 12 = 232$$

and so

$$\bar{x} = \frac{232}{15} = 15.5$$

The mean price of an orange is 15.5 pence. Do not be put off by the fact that it is impossible for an individual orange to cost 15.5p, this is still useful information about the price of an orange.

In the above example the prices of the oranges might have been recorded in a *frequency table,* as in Table 9.2. This table shows, for example, that 5 out of the 15 shops sold their oranges at 15 pence each.

Table 9.2.
Frequency table of orange prices

Price per orange	Number of shops
11	1
12	2
13	1
15	5
17	2
18	2
19	1
20	1
Total	15

From the definition of the mean,

$$\bar{x} = \frac{\text{total of orange prices}}{\text{number of oranges}}$$

From the table, the number of oranges is 15. The total of the prices for these oranges is

$$1(11) + 2(12) + 1(13) + 5(15) + 2(17) + 2(18) + 1(19) + 1(20) = 232$$

$$\text{The mean} = \frac{232}{15} = 15.5 \text{ pence}$$

as before.

KEY POINT

The formula for calculating the mean from grouped data is

$$\bar{x} = \frac{\sum xf}{\sum f}$$

where $\sum xf$ is the sum of the value × frequency column
and $\sum f$ is the sum of the frequency column.

Worked example

9.2   Table 9.3 shows the daily demand for hire cars belonging to White Cars Ltd. Find the mean daily car demand.

Table 9.3.
Demand for cars

Daily demand	Frequency
1	3
2	4
4	7
5	12
6	12
8	7
9	4
10	1

Solution   The relevant calculations are shown in the following table:

x	f	xf
1	3	3
2	4	8
4	7	28
5	12	60
6	12	72
8	7	56
9	4	36
10	1	10
	50	273

$$\bar{x} = \frac{\sum xf}{\sum f} = \frac{273}{50} = 5.46$$

The mean demand is 5.46 cars per day.

The use of a frequency table is of particular use when there is a large volume of data. It reduces the amount of calculation in finding the mean.

A similar process takes place when the data are not available in raw form but already compiled in a frequency table. Table 9.4 shows the times taken by maintenance staff to repair a particular type of equipment fault. We have seen earlier that to calculate a mean two pieces of information are needed (here, the number of repairs and the total time taken to carry out these repairs). The first of these is straightforward, since we can total the frequency column to obtain 250. But how do we find out the total time taken to carry out these repairs?

Table 9.4.
Times to repair
equipment

Time taken (in minutes)	Frequency
0 and under 10	28
10 and under 20	42
20 and under 30	56
30 and under 40	74
40 and under 50	32
50 and under 60	18

The first class consists of the 28 repairs that take up to 10 minutes. Without any extra information we assume that all the times in the class fall at the centre of the class; that is, the midpoint of 0 and 10, which is 5 minutes. This value would also be appropriate when the values are spread evenly through the class. This is repeated for all the classes. For example in the second class we assume that all 42 times take the value of 15 minutes. Now we proceed as before by multiplying the midpoint by the frequency to give an estimate for the total time taken by the repairs falling in that class. When totalled for all classes we have a good estimate for the total time taken to carry out all the repairs. The mean is then obtained by dividing by the overall frequency. The procedure is normally set out using a table in the following way; see Table 9.5.

Table 9.5.
Calculations for mean
repair time

Class (min)	Midpoint, x	Frequency, f	Repair time × frequency
0–10	5	28	140
10–20	15	42	630
20–30	25	56	1400
30–40	35	74	2590
40–50	45	32	1440
50–60	55	18	990
Total		250	7190

$$\text{Mean repair time} = \frac{7190}{250} = 28.76 \text{ minutes}$$

Worked example

9.3   The value of 100 invoices issued by a firm on one day is shown in Table 9.6. Find the mean value per invoice.

Table 9.6.
Value of 100 invoices

Value (£)	Frequency
0 and under 5	2
5 and under 10	12
10 and under 15	29
15 and under 20	22
20 and under 25	18
25 and under 30	10
30 and under 40	5
40 and under 50	2

Solution   The calculations are shown in Table 9.7.

Table 9.7.
Calculations for mean
invoice table

Class	Midpoint, x	Frequency, f	xf
0–5	2.5	2	5
5–10	7.5	12	90
10–15	12.5	29	362.5
15–20	17.5	22	385
20–25	22.5	18	405
25–30	27.5	10	275
30–40	35	5	175
40–50	45	2	90
	Total	100	1787.5

$$\bar{x} = \frac{1787.5}{100} = £17.88$$

The mean invoice value is £17.88.

You will note in the calculations shown in Table 9.7 that the first class was listed as going from 0 to £5, rather than the correct range from 0 to £4.99. This latter (more accurate) range would have given a midpoint of 2.495 and an xf value of 4.99. The approximation is repeated in all other classes. Had we used the accurate values in all classes we would have got virtually the same answer (in fact, $\bar{x} = £17.87$) but it would have involved more demanding arithmetic and increased the likelihood of an error. In general, it is sensible when calculating the mean to round the class boundaries as long as it still gives the appropriate degree of accuracy.

Self-assessment question 9.2

1. Why might there be a difference between a mean calculated from raw data and one calculated from grouped data?

Exercise 9.2

1. Five workers earn £180, £195, £195, £200 and £220 a week. What is their average wage? What must a sixth worker earn if the average of the six is to be £1 more?

2. Four children are aged 11 years, 11 years 1 month, 11 years 3 months, 11 years 4 months. Find their average age.

3. A car travels for 2 hours at 25 miles per hour and 90 minutes at 30 miles per hour. Find the total time, total distance and average speed.

4. Find the average wage for the following group of people described in Table 9.8.

Table 9.8. Wage distribution

Wage (£)	Frequency
172	15
175	20
180	40
185	35
190	17
192	13
195	10

5. The average monthly earnings of 200 employees working for Tritown Ltd are shown in Table 9.9. Find the mean monthly earnings for these 200 employees.

Table 9.9. Monthly earnings

Monthly earnings (£)	Number of employees
400 and under 600	24
600 and under 800	50
800 and under 1000	64
1000 and under 1200	42
1200 and under 1400	20

9.3   Median

The second type of average is the *median*, or middle, value where half of the numbers are less than or equal to the median and half are more than or equal to the median.

> I'd be surprised if this was not true?

KEY POINT

To find the median of a set of numbers, arrange them in ascending order of size and pick out the one in the middle.

For example,

2, 4, 9, 7, 5

has median 5, for arranged in order they are 2, 4, 5, 7, 9 and 5 is in the middle. Another example is

1, 2, 4, 5, 6, 6, 8, 9, 10

which has median 6. These are already in ascending order and the value of the one in the middle is 6. It does not matter that there is another 6 as well.

In both of these examples there have been an odd number of items. If the number of items is even, the definition given at the beginning has to be modified slightly. Then the values are arranged in ascending order and the two in the middle are picked out. The median is then found by computing the mean of these two numbers. For example,

11, 8, 6, 2, 3, 12, 10, 14

has median 9, for arranged in order they are 2, 3, 6, 8, 10, 11, 12, 14; the two in the middle are 8 and 10, and these two numbers have mean 9.

Worked example

9.4   Table 9.10 shows the number of children in 30 households. Determine the median number of children per household.

Table 9.10.
Number of children in 30 households

2	1	0	7	4	1	2	0	2	3
3	4	2	1	2	5	1	3	2	4
2	1	0	2	3	3	1	2	3	2

Solution   In ascending order the data are

0, 0, 0, 1, 1, 1, 1, 1, 1, 2, 2, 2, 2, 2, 2, 2, 2, 2, 2, 3, 3, 3, 3, 3, 3, 4, 4, 4, 5, 7

There are 30 numbers, so the median is given by the mean of the 15th and 16th numbers. Counting from the left the 15th number is 2, as is the 16th number.

Median = 2

thus indicating that an average or typical family has 2 children.

If there are a large number of items it may be easier to find the median by alternatively striking out the largest and smallest numbers. Using this last example, first strike out the largest in the set (7), then strike out the smallest (0), next strike out the second largest (5), then strike out the second smallest (0). Continue in this way, alternately striking out the largest remaining value and the smallest remaining value until one number (or two numbers if the number of items is even) is left.

2	1	0	~~7~~	4	1	2	~~0~~	2	3
3	4	2	1	2	~~5~~	1	3	2	4
2	1	~~0~~	2	3	3	1	2	3	2

Worked example

9.5 The daily demand for a product is given in Table 9.11. State the median.

Table 9.11.
Product demand

Demand	Frequency
2	4
3	6
4	4
5	7
6	2
7	4

Solution Writing the 27 values out in order gives

2, 2, 2, 2, 3, 3, 3, 3, 3, 3, 4, 4, 4, 4, 5, 5, 5, 5, 5, 5, 5, 6, 6, 7, 7, 7, 7

The middle (14th) value is 4, so the median is 4.

Worked Example

9.6 The incomes of six bank employees are

£15 000   £17 000   £12 500   £14 500   £19 000   £18 000

Find   (i) the mean,
       (ii) the median.

Solution (i) mean $= \dfrac{\text{total}}{6} = \dfrac{96\,000}{6} = £16\,000$

(ii) median $= (15\,000 + 17\,000)/2 = £16\,000$

Referring to worked example 9.6, imagine now that the bank employee earning £19 000 earns not £19 000 but £39 000. What happens to the mean and median?

The median does not change, but the mean increases to £19 333. It is true that the mean is now unrepresentative because it is significantly influenced by the extreme value.

Self-assessment questions 9.3

1. Describe one advantage and one disadvantage of using the median as a typical value, compared with using the mean.

2. Comment on the statement 'Half of the population have below average intelligence'.

Exercise 9.3

1. The incomes of seven employees are

   £14 000   £16 000   £18 000   £11 000   £19 000   £21 000   £18 000

   Find the median income. Imagine an eighth employee joins the team and earns an income of £15 000. What is the new value of the median?

2. The frequency table below indicates the ages of 24 children. What is the median age?

Age	Frequency
4	2
5	8
6	9
7	4
8	1

3. The maximum temperatures (in °C) on 30 consecutive days of a given month are given below:

   8 12 6 7 9 11 12 10 14 9 6 3 4 3 7 9 11 11 12 13 10 9 7 6 8 9 6 5 7 9

   Find the median for the data.

9.4	**Mode**	

The third type of average is the *mode* which is used to indicate the value (or type) that occurs most often. The mode is an 'average' in the sense that it is the most common.

KEY POINT

> The mode is found by counting the number of times each value occurs and then selecting the value with the highest frequency.

For example, the numbers

   4, 8, 12, 10, 7, 5, 8, 14, 20

have a mode equal to 8 because it occurs twice and no other number occurs that often.

Worked example

9.7 A worker at a post office keeps a record of the postage stamps sold at his branch. In one day 160 postage stamps were sold to its customers. The cost and frequency of each is shown in Table 9.12. Determine the mode.

Table 9.12.
Type and frequency of postage stamps

Type of postage stamp	Frequency
20p	45
26p	64
31p	10
37p	4
44p	2
50p	7
100p	18

Solution   The postage stamp most frequently bought is that which costs 26p. This is the mode.

We have seen how the mean, median and mode can be obtained from a set of numbers. All three of these measures can be used to indicate the 'average' of the numbers in their different meanings. Consequently the word 'average' might lead to confusion if the type of average is not indicated. Consider the following example.

Worked example

9.8 The annual income of 20 households in one street is shown in Table 9.13. In each case the annual income is recorded to the nearest £100.

Table 9.13.
Annual incomes (£)

5 400	6 000	6 000	6 300	6 600
7 200	7 500	9 000	9 000	9 000
9 600	10 200	10 800	12 000	12 000
13 500	14 400	15 000	18 000	22 500

Determine (a) the mean, (b) the median and (c) the mode.

Solution   (a) The total of all annual incomes is

$$\sum x = 5400 + 6000 + 6000 + \ldots + 22\,500 = 210\,000$$

$$\text{The mean income} = \frac{210\,000}{20} = £10\,500$$

(b) The incomes are already in ascending order.

$$\text{The median income} = \frac{9000 + 9600}{2} = £9300$$

(c) The income that occurs most frequently, and hence the mode, is £9000.

These three values are all different, emphasizing the need for clarity when using the term 'average'.

Self-assessment question 9.4

1. Describe a situation where the mode might be used in preference to the median and mean as a typical value.

Exercise 9.4

1. Find the mode of the following set of data:

     5 4 2 7 3 4 6 2 9 8 2 4 1 3 5 7 4 9 8 6 4 2 9 8 7 6 6 4 3 4

2. The table below indicates the colour of car passing a particular point in a one-hour interval. Which colour was most common?

Colour	Frequency
Blue	32
Red	51
Black	8
White	15
Green	20

3. A shop sells two pairs of shoes size 6, seven pairs of size 8, eight pairs of size 9, four pairs of size 10, and one pair of size 11. State the modal size of shoe sold.

Test and assignment exercises 9

1. The 1996 temperature, maximum and minimum, and sunshine figures for England and Wales are available for each month, and are shown in Table 9.14.

Table 9.14. Temperature and sunshine for 1996 in England and Wales

Months in 1996	Maximum air temp. (°C)	Minimum air temp. (°C)	Mean daily sunshine (hours)
January	14.3	−11.0	1.64
February	14.5	−5.8	2.60
March	18.8	−8.6	5.23
April	22.1	−5.1	5.89
May	28.6	−4.3	6.91
June	33.0	0.3	6.80
July	32.7	1.8	7.44
August	34.6	1.5	8.82
September	23.2	−3.2	4.65
October	25.6	−2.6	4.04
November	16.8	−6.2	2.26
December	14.5	−16.4	1.37

Source: Monthly Digest of Statistics, 1997. Office for National Statistics. Crown Copyright, 1997.

Assuming that each month has equal weight determine
(a) the mean maximum air temperature,
(b) the mean minimum air temperature,
(c) the mean daily sunshine for 1996 as a whole.

2. A footwear shop sells sports shoes in whole sizes only. One summer, 260 sports shoes were sold at this shop. A breakdown of the shoe sizes sold can be seen in Table 9.15. What is the modal size of shoe?

Table 9.15. Frequency table of shoe sizes

Shoe size	Frequency
3	8
4	12
5	25
6	43
7	65
8	57
9	32
10	18

3. One cricket season Alan Smith batted on 18 occasions. His scores were

9, 32, 23, 0, 56, 28, 4, 12, 66, 84, 59, 3, 0, 42, 19, 8, 22, 26

Determine the mean score.

4. Brian travelled to work by car. His journey times, in minutes, over a three-week period are shown in Table 9.16. Determine the median

Table 9.16. Journey times (minutes)

	Monday	Tuesday	Wednesday	Thursday	Friday
Week 1	42	38	39	36	47
Week 2	43	41	35	38	44
Week 3	41	36	37	39	42

5. Eight different tour operations organize a seven-day holiday to a certain hotel in a Spanish holiday resort. The quoted price of each of these holidays is, in pounds,

384, 396, 354, 387, 389, 399, 355, 360

Determine the mean price of a tour to this hotel.

6. At eight performances of a play in a week the numbers present were

873, 681, 752, 942, 621, 826, 1036, 1092

Determine the median attendance per performance.

7. Petrol prices (per litre) at twelve different garages in a given town centre were

67.9p, 65.5p, 64.9p, 69.9p, 68.2p, 67.9p 68.9p, 66.3p, 67.5p, 67.9p, 66.6p, 68.7p

Determine (a) the mean, (b) the median and (c) the mode.
Which of these averages do you think is the most appropriate?

8. For the data in Table 9.16 calculate the mean and compare with the median.

10 Spread

Objectives

After reading this chapter, you should be able:

- to describe why measures of spread are needed
- to calculate the range and standard deviation for individual data
- to calculate the coefficient of variation
- to explain the meaning of the standard deviation and the coefficient of variation

10.1 Introduction

In the last chapter three measures of centrality or average, the mean, median and mode, were used to describe and summarize a set of data. It is often necessary to supplement a statement about the position of the average of the data set by some indication of how closely the data are concentrated about the average.

Suppose that we want to compare the average weekly earnings (£) of sales representatives in two companies, Apex and Bemax. Table 10.1 shows the earnings for ten sales representatives in each company.

Table 10.1.
Earnings of sales
representatives

| Apex | 230 | 240 | 240 | 260 | 220 | 250 | 280 | 250 | 260 | 270 |
| Bemax | 230 | 220 | 280 | 150 | 200 | 300 | 240 | 270 | 260 | 350 |

The mean can be calculated in each case to be £250. However, the weekly incomes for Apex are much more concentrated about the mean. Therefore the mean of £250 is more typical of the weekly earnings for Apex. In other words, the mean of Bemax earnings is less reliable than that of Apex as an indication of a typical value. Figure 10.1 illustrates the differences in variability between the two companies.

Figure 10.1.
Earnings distributions
for two companies

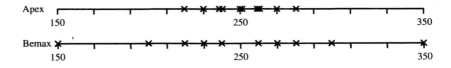

One of the simplest measures of spread is the range, which is the difference between the largest value and the smallest.

> The range is the difference between the largest observed value and the smallest.

For the Apex company it can be seen that the largest observed value is £280 and the smallest is £220. The ten weekly wages range between £220 and £280. In statistical terms

Range = 280 − 220 = £60

In the Bemax company

Range = 350 − 150 = £200

This illustrates that the weekly earnings of the Bemax employees are more spread out compared with the other company.

Worked example

10.1 Eight students are given their coursework and examination marks (%) for their mathematics module at University. The results are recorded in Table 10.2.

Table 10.2.
Coursework and examination results

Student	Coursework mark (%)	Examination mark (%)
A	70	50
B	63	80
C	74	54
D	65	45
E	59	59
F	57	47
G	65	64
H	56	77

Use the range to identify which of the two sets of marks are more spread out.

Solution For coursework marks, the range is 74 − 56 = 18

For examination marks, the range is 80 − 45 = 35

It is clear that the examination marks are more spread out than the coursework marks.

It is seen that the calculation of the range is easy but this measure uses only the extreme values within the data set. Small changes to other values will not affect the range. For this reason, the range tends not to be used in practical statistical analysis. An important exception to this last point is in small sample quality control, where use is made of small random samples of constant size (typically four to five items) to identify whether some measured dimension is under control.

10.2   Standard deviation

The previous section illustrated the need for a measure of spread for data, but indicated that the range is not always a suitable measure. The most common measure of spread in statistics is the standard deviation. The standard deviation uses all the available data to measure the variation within a set of data and is based on the differences of the data values from the mean. The more spread out the data, the greater the differences are and the higher is the measure of spread. Thus a low standard deviation indicates that the data are concentrated tightly around the mean.

KEY POINT

The standard deviation for a set of data is defined as
$$s = \sqrt{[\sum (x - \bar{x})^2 / n]}$$

To use the formula, we must first calculate the mean and then for each data value:

1. Subtract the mean, $\bar{x}$, to give $x - \bar{x}$.
2. Square $(x - \bar{x})$ to give $(x - \bar{x})^2$.

We then add together the values of $(x - \bar{x})^2$ for each data value and divide by n, the number of data values. Finally, we take the square root to give the standard deviation.

A table is often used to help to perform the calculation. We use an example to illustrate the calculation.

Worked example

10.2 Two friends were comparing their assessment marks at the end of the year. Ben and Bev's marks are shown in Table 10.3. For each friend, calculate the mean and standard deviation and compare their performances.

Table 10.3.
Module marks for two students

Subject	Ben	Bev
Mathematics	62	84
Computing	59	40
Business	75	72
Study skills	56	56
Social studies	70	50
Accounting	68	64

Solution To distinguish between the two sets of marks, we represent Ben's marks by x and Bev's marks by y.

Taking Ben first, the mean of Ben's marks is

$$\bar{x} = (62 + 59 + 75 + 56 + 70 + 68)/6 = 65$$

The first data item is 62. Therefore we have

$$(x - \bar{x}) = 62 - 65 = -3$$

which gives us

$$(x - \bar{x})^2 = (-3)^2 = (-3) \times (-3) = 9$$

This process is repeated for each data item. Table 10.4 shows the calculations involved.

Table 10.4.

x	$\bar{x}$	$(x - \bar{x})$	$(x - \bar{x})^2$
62	65	−3	9
59	65	−6	36
75	65	10	100
56	65	−9	81
70	65	5	25
68	65	3	9
			Total = 260

The standard deviation for Ben's marks is

$$s = \sqrt{(260/6)} = \sqrt{43.33} = 6.58$$

For Bev, the mean assessment score is

$$\bar{y} = (84 + 40 + 72 + 56 + 50 + 64)/6 = 61$$

Table 10.5 gives the calculations.

Table 10.5.

y	$\bar{y}$	$(y - \bar{y})$	$(y - \bar{y})^2$
84	61	23	529
40	61	−21	441
72	61	11	121
56	61	−5	25
50	61	−11	121
64	61	3	9
		Total =	1246

The standard deviation for Bev's marks is

$$s = \sqrt{(1246/6)} = \sqrt{(207.67)} = 14.41$$

It can clearly be seen from the means that, on average, Ben has higher marks than Bev. Also Ben's marks are less spread out; that is, they are more concentrated around the mean score of 65. Bev's marks are less consistent (or more erratic) .

The standard deviation is in practice a very important measure of spread because of its mathematical properties, which allow valuable results to be readily deduced. The mean and standard deviation together provide a powerful summary of a set of data.

Note that the standard deviation is never a negative number. This is because $\sum (x - \bar{x})^2$ is a sum of terms which is always greater than or equal to zero.

Worked example

10.3   The mean temperature of 10 hospital patients were measured. The results are as follows (in °C)

38.0   37.7   38.6   37.0   36.8   37.4   38.8   37.5   37.8   38.4

Find the mean and standard deviation.

Solution   The mean temperature is

$$\bar{x} = (38.0 + 37.7 + \ldots + 38.4)/10 = 37.8$$

Calculations are shown in Table 10.6, from which the standard deviation can be obtained as follows:.

$$s = \sqrt{(3.94/10)} = \sqrt{0.394} = 0.63$$

Table 10.6.

x	$\bar{x}$	$x - \bar{x}$	$(x - \bar{x})^2$
38.0	37.8	0.2	0.04
37.7	37.8	−0.1	0.01
38.6	37.8	0.8	0.64
37.0	37.8	−0.8	0.64
36.8	37.8	−1.0	1.00
37.4	37.8	−0.4	0.16
38.8	37.8	1.0	1.00
37.5	37.8	−0.3	0.09
37.8	37.8	0.0	0.00
38.4	37.8	0.6	0.36
		Total =	3.94

It is useful to know that there are different formulae for finding the standard deviation of a set of data. A well used formula, which gives the same answer as that given by the current approach, is described in the next key point. This formula is generally easier and quicker to use, particularly when the mean is not an easy-to-use whole number.

KEY POINT

An alternative formula for finding the standard deviation of a set of data is

$$s = \sqrt{\left(\sum x^2 / n - \bar{x}^2\right)}$$

where $\sum x^2$ is the sum of squares of all the numbers in the data set.

For the data of worked example 10.3 the mean is $\bar{x} = 37.8$, and

$$\sum x^2 = 38.0^2 + 37.7^2 + 38.6^2 + 37.0^2 + 36.8^2 + 37.4^2 + 38.8^2$$
$$+37.5^2 + 37.8^2 + 38.4^2 = 14\,292.34$$

Consequently, the standard deviation is

$$s = \sqrt{(14\,292.34/10 - 37.8^2)}$$
$$= \sqrt{(1429.234 - 1428.84)} = \sqrt{0.394} = 0.63$$

which agrees with the answer given by the alternative approach. It is particularly important to keep as much accuracy as possible when using this latter approach for calculating the standard deviation.

Looking at worked example 10.2 again, this latter approach also gives the same answers as those given earlier.

Here the data set for Ben was 62, 59, 75, 56, 70, 68 with a mean of 65. The standard deviation could have been calculated using

$$\sum x^2 = 62^2 + 59^2 + 75^2 + 56^2 + 70^2 + 68^2 = 25\,610$$
$$s = \sqrt{(25\,610/6 - 65^2)} = \sqrt{(4268.333\,333 - 4225)}$$
$$= \sqrt{43.333\,333} = 6.58$$

For Bev the data set was 84, 40, 72, 56, 50, 64 giving a mean of 61 and $\sum y^2 = 23\,572$, giving

$$s = \sqrt{(23\,572/6 - 61^2)} = \sqrt{(3928.666\,667 - 3721)}$$
$$= \sqrt{207.666\,667} = 14.41$$

Self-assessment questions 10.2

1. Explain why the standard deviation is a more reliable measure of spead than the range.

2. The times, measured in minutes, taken by six people to have a shower are

    8.2   6.8   12.1   4.7   11.5   10.7

    Calculate the standard deviation.

Exercise 10.2

1. Five workers earn £80, £95, £95, £100, £120 per week. Calculate the mean and standard deviation of these wages.

2. In two consecutive seasons Michael Brown's cricket scores were

Season 1:	9	32	23	0	56	28	4	16		
Season 2:	66	84	59	3	0	42	19	9	22	26

    Calculate the mean and standard deviation of the scores for each season. Interpret these values.

3. Graham travelled to work by car. His journey times (in minutes) on Wednesdays and Fridays over a six-week period were

Wednesdays	39	35	37	38	41	38
Fridays	46	44	42	42	43	41

    For each of the two days calculate the mean and standard deviation.

4. Petrol prices (pence per litre) at eight different garages in a given town centre were

    62.8   64.9   65.3   61.7   66.2   64.5   63.9   65.9

    Calculate the mean price and standard deviation.

5. A quality control inspector found the following number of defective parts on eight different days on an assembly line production

    3   10   8   9   6   10   12   6

    Calculate the mean and standard deviation.

10.3   Coefficient of variation

We often need to consider the variation in two sets of data whose items are measured in different units, or are of different sizes. Suppose that we are asked to compare the variability of sets of salaries; for example, the weekly earnings of a group of workers in the United Kingdom and a group of workers in the same profession in the United States. The values of the standard deviation may not be meaningful even if the data are measured on a common scale (e.g. pounds or dollars). If the two data sets have mean values of £120 and £250 respectively, and they both have a standard deviation of £25, then in relative terms the spread of the second data set is less than that of the first.

We are able to compare the spread (or variability) of the data set by considering the spread in the data relative to the mean. The coefficient of variation, CV, is defined as the standard deviation divided by the mean, and is often expressed as a percentage.

KEY POINT

> The coefficient of variation is based on the mean and standard deviation and is calculated as
>
> $$CV = \frac{\text{standard deviation}}{\text{mean}} \times 100\%$$

Worked example

10.4   We wish to compare the day-to-day variabilities in mileage and petrol consumption for a travelling salesman. The mileages per day have a mean of 200 miles with a standard deviation of 30 miles, while the petrol consumption has a daily mean consumption of 25 litres and a standard deviation of 6 litres. Calculate the coefficient of variation for both mileage and petrol consumption.

Solution   With the given information it is difficult to compare the variability in mileage and petrol consumption. We can calculate the coefficients of variation to aid the comparison.

For mileage:

$$\text{coefficient of variation} = \frac{30}{200} \times 100\% = 15\%$$

that is, the standard deviation is 15% of the mean.

For petrol consumption:

$$\text{coefficient of variation} = \frac{6}{25} \times 100\% = 24\%$$

This indicates a greater variability in petrol consumption than in mileage.

Worked example

10.5 A small engineering firm has recorded the bonuses paid to its nine workshop staff:

$$£17 \quad £12 \quad £24 \quad £8 \quad £10 \quad £20 \quad £25 \quad £16 \quad £30$$

Determine the mean, standard deviation and coefficient of variation.

Solution $n = 9$ and the mean is $162/9 = £18$. If x represents bonus payments,

$$\sum x^2 = 17^2 + 12^2 + 24^2 + 8^2 + 10^2 + 20^2 + 25^2 + 16^2 + 30^2 = 3354$$

The standard deviation is

$$\sqrt{(3354/9 - 18^2)} = \sqrt{(372.6667 - 324)}$$
$$= \sqrt{48.6667} = £6.98$$

The coefficient of variation is $\dfrac{6.98}{18} \times 100\% = 39\%$

Self-assessment question 10.3

1. The shop prices (in pence) of apples and oranges were obtained from 15 greengrocer shops on one Saturday, and shown in Table 10.7.

Table 10.7. Shop prices for apples and oranges

Shop	A	B	C	D	E	F	G	H	I	J	K	L	M	N	O
Apples (1 kg)	75	80	86	90	74	78	69	82	79	94	79	85	94	70	80
Oranges (each)	17	15	18	20	15	15	11	15	19	18	15	17	12	13	12

Explain why the coefficient of variation is an appropriate measure to use as a comparison of spread.

Calculate the standard deviation and the coefficient of variation for each fruit to identify which of the sets of prices is the more variable.

Exercise 10.3

1. The earnings (£ per week) of ten employees of Department A are

$$237 \quad 245 \quad 283 \quad 296 \quad 253 \quad 249 \quad 236 \quad 254 \quad 305 \quad 242$$

Given that the employees of Department B have a mean weekly wage of £366 and a standard deviation of £29, compare the earnings of the two departments.

2. A company has two factories situated in two different parts of the country. Managerial turnover in Factory Y is much higher than in Factory X, and a possible cause of this is the greater inequality of income in Factory Y. The following data are collected.

Factory	Mean income	Standard deviation
X	£15 000	£5 000
Y	£20 000	£15 000

Calculate the coefficient of variation for each factory. What would you conclude?

3. The heights (in cm) of five people are recorded below

181   169   179   175   171

Calculate the mean, standard deviation and coefficient of variation.

Test and assignment exercises 10

1. Pollen counts were taken in regions of equal area, where two different plant treatments had been used. Four regions received Treatment 1, eight received Treatment 2. The pollen counts are given below.

Treatment 1	77	61	157	52				
Treatment 2	17	31	87	16	18	26	77	20

Use (i) the range and (ii) the standard deviation to compare the variabilities of the pollen counts for the two treatments. Which of these two measures is preferable in making this comparison?

2. The 1996 temperature, maximum and minimum, and sunshine figures for England and Wales are available for each month, and are shown in Table 10.8.

Calculate the coefficient of variation for each of the three sets of figures to identify which is the least and which the most variable.

Table 10.8. Temperature, rainfall, and sunshine for 1996 in England and Wales

Months in 1996	Maximum air temp. (°C)	Minimum air temp. (°C)	Mean daily sunshine (hours)
January	14.3	−11.0	1.64
February	14.5	−5.8	2.60
March	18.8	−8.6	5.23
April	22.1	−5.1	5.89
May	28.6	−4.3	6.91
June	33.0	0.3	6.80
July	32.7	1.8	7.44
August	34.6	1.5	8.82
September	23.2	−3.2	4.65
October	25.6	−2.6	4.04
November	16.8	−6.2	2.26
December	14.5	−16.4	1.37

Source: Monthly Digest of Statistics, Nov. 1997. Office for National Statistics, Crown Copyright.

3. The times (in minutes) taken to repair machines of two types are recorded below:

Machine type A:	82	90	100	85	95
Machine type B:	24	30	15	32	40

Calculate the mean, standard deviation and coefficient of variation to compare the repair times of the machine types.

11 Correlation

Objectives

After reading this chapter, you should be able:

- to construct and interpret a scatter diagram
- to calculate the product moment correlation coefficient
- to assess the correlation coefficient

11.1 Introduction

In the previous two chapters we investigated ways in which one set of data can be described and summarized. The mean, for example, describes the centre of the data, in some way, and the standard deviation describes the spread of the data. In this chapter we measure how strong the relationship is between two sets of data; as for example in Table 11.1. In each of the three situations there is likely to be some relationship between the first variable and the second variable. The aim of this chapter is to provide some idea on the strength of the relationship.

Table 11.1.
Possible relationships between two sets of data

	Variable 1	Variable 2
1.	Heights of people	Weights of people
2.	Students' coursework marks	Students' examination marks
3.	Ages of plants	Quantities of fruit produced by the plants

Is this positive correlation?

11.2   Scatter diagrams

Suppose we consider the coursework marks and examination marks for ten students on the same course at university, as shown in Table 11.2. Why might we be interested in identifying if there is a relationship between coursework and examination marks? If there is a relationship then a student may have a good idea about his or her prospective examination performance from the already attained coursework performance.

Table 11.2.
Coursework and examination marks for ten students

Student	Coursework mark (%)	Examination mark (%)
A	52	50
B	60	62
C	75	70
D	40	50
E	50	55
F	48	50
G	55	58
H	70	70
I	60	40
J	80	75

One useful first step would be to show the data in the form of a diagram or chart – often known as a scatter diagram. The construction of such a diagram follows from the principles studied in Chapter 7 on co-ordinates. In Figure 11.1 each pair of values from the two variables is plotted as a single point to see whether there is a pattern among the points.

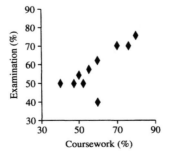

Figure 11.1.
Scatter diagram showing the marks for ten students

You will see in Figure 11.1 that we have put coursework marks on the x axis and examination marks on the y axis. For each student (that is, for each pair of coursework marks and examination marks) we simply place that student at the suitable co-ordinates on the graph. Check that you can locate the point for each student on the graph.

There are a number of things we can say to interpret the relationship between the two variables:

1. In general, the higher the coursework mark, then the higher the examination mark.
2. The scatter of points represents a reasonably close pattern running from bottom left to top right.
3. While the pattern of points is reasonably close, it is not perfectly so.

> Correlation provides a measure of the strength of relationship between two variables.

As the points are close to a straight line we can report that there is *strong* linear correlation between coursework marks and examination marks. Furthermore, as high coursework marks are associated with high examination marks we can report that there is strong *positive* correlation between coursework and examination marks. Note that it is possible to identify non-linear relationships, but this chapter will only consider the measurement of linear correlation.

Worked example

11.1   A sample of eight employees is taken from the production department of a light engineering factory. The data in Table 11.3 relate to the number of weeks' experience in the wiring of components, and the number of components which were rejected as unsatisfactory last week.

Table 11.3.

	Employee							
	A	B	C	D	E	F	G	H
Weeks of experience (x)	4	5	7	9	10	11	12	14
Number of rejects (y)	21	22	15	18	14	14	11	13

Draw a scatter diagram and report verbally on the relationship.

Solution   From these results we are able to draw the scatter diagram of Figure 11.2.

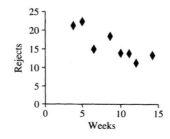

Figure 11.2.
Scatter diagram for number of rejects against weeks of experience

Note that we have put the number of weeks of experience on the x axis and the number of rejects on the y axis. In general, as stated in Chapter 8, we select x as the independent variable and y as the dependent variable. The linkage between the two variables – which is dependent (y) and which is independent (x) – is based on what we believe to be the correct relationship between the variables.

From Figure 11.2 we see that there is a strong negative correlation between the two variables: strong because the points are close to a straight line, negative because a high number of weeks' experience is associated with a low number of rejects.

Clearly, there are a large number of possible scatter diagrams, resulting from different data sets, as shown in Figures 11.3–11.5.

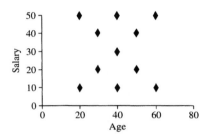

Figure 11.3.
No correlation

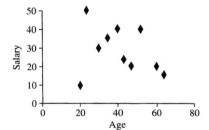

Figure 11.4.
Weak negative correlation

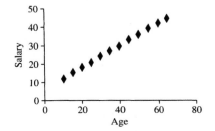

Figure 11.5.
Perfect positive
correlation

Self-assessment question 11.2

1. A government-financed Industrial Research Unit selected ten factories at random from all those engaged in heavy engineering in this country. They were similar in all respects except size of workforce. Each factory was asked to indicate the percentage of employees absent from work for at least one whole day during a particular month. The results are tabulated in Table 11.4. Draw a scatter diagram and comment on the type of correlation.

Table 11.4. Workforce size and percentage of absentees

Factory	Workforce size (000)	Percentage of absentees
A	1.1	5
B	3.0	6
C	5.1	8
D	5.8	9
E	7.0	11
F	8.3	12
G	9.3	14
H	10.0	20
I	10.9	43
J	12.1	60

Exercise 11.2

1. Figures for black and white television licences and colour television licences are given in Table 11.5. Draw a scatter diagram and interpret it.

Table 11.5. Black and white television licences and colour television licences

Year	Black and white (millions)	Colour (millions)
1988	2.01	17.31
1989	1.76	17.85
1990	1.52	18.09
1991	1.26	18.15
1992	1.07	18.71
1993	0.93	19.25
1994	0.80	19.67
1995	0.66	20.28

Source: Monthly Digest of Statistics, 1996. Office for National Statistics. Crown Copyright.

2. Two friends, Ben and Bev, were comparing their assessment marks at the end of the year. Their marks are shown in Table 11.6. Construct a scatter diagram and interpret it.

Table 11.6. Assessment marks

Subject	Ben	Bev
Mathematics	62	84
Computing	59	40
Business	75	72
Study skills	56	56
Social studies	70	50
Accounting	68	64

3. Coursework and examination marks for eight students are available for a given module (Table 11.7). Draw a scatter diagram and comment on it.

Table 11.7. Module results

Student	Coursework mark (%)	Examination mark (%)
A	70	50
B	63	80
C	74	54
D	65	45
E	59	59
F	57	47
G	65	64
H	56	77

11.3   ## Correlation coefficient

The degree of correlation between two variables can be measured, and we can decide, using actual data, whether two variables are perfectly, strongly or only weakly correlated, and whether the correlation is positive or negative. This degree of correlation is measured by the correlation coefficient (sometimes called Pearson's correlation coefficient or the product–moment correlation coefficient). In fact, there are different ways of defining a correlation coefficient, but Pearson's correlation coefficient is the most frequently used when both sets of data are numerical values and it is the only one covered in this chapter.

As with the standard deviation there are a number of different formulae for a set of data that can give the same value for this correlation coefficient. The one that is most commonly used, and probably the easiest to use, is given in the next key point.

KEY POINT

The correlation between two sets of data can be measured by

$$\text{Correlation coefficient, } r = \frac{n \sum xy - \sum x \sum y}{\sqrt{\{[n \sum x^2 - (\sum x)^2][n \sum y^2 - (\sum y)^2]\}}}$$

where x and y represent pairs of data for the two variables and n is the number of pairs of data.

Note that

- $\sum x$ is obtained by adding all values of x,
- $\sum y$ is obtained by adding all values of y,
- $\sum x^2$ is obtained by squaring each value of x and totalling for all values,
- $\sum y^2$ is obtained by squaring each value of y and totalling for all values,
- $\sum xy$ is obtained by multiplying each value of x by its corresponding y-value and totalling for all values.

These summations are usually obtained by putting the data in a table.

The value of the correlation coefficient, r, always falls between -1 and $+1$. If you get a value outside this range you have made a mistake. The closer the value is to $+1$ or -1 the stronger the degree of correlation. Figures 11.6–11.9 illustrate the correlation value for various data sets.

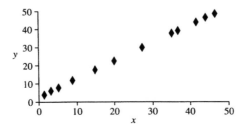

Figure 11.6.
Scatter diagram for
$r = +0.99$

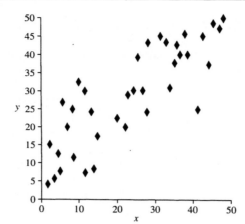

Figure 11.7.
Scatter diagram for
$r = +0.8$

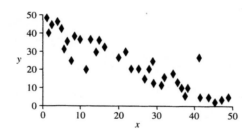

Figure 11.8.
Scatter diagram for
$r = -0.9$

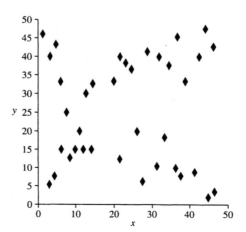

Figure 11.9.
Scatter diagram for
$r = 0.02$

KEY POINT

As a rule of thumb the correlation coefficient can be interpreted in the following way:

Value of r	Interpretation
$r = +1$	Perfect positive correlation
r lies between $+0.9$ and $+1$	Very strong positive correlation
r lies between $+0.7$ and $+0.9$	Good positive correlation
r lies between $+0.3$ and $+0.7$	Weak positive correlation
r lies between -0.3 and $+0.3$	Virtually no correlation
r lies between -0.3 and -0.7	Weak negative correlation
r lies between -0.7 and -0.9	Good negative correlation
r lies between -0.9 and -1.0	Very strong negative correlation
$r = -1.0$	Perfect negative correlation

Remember, correlation can be positive or negative. Positive correlation means that low values of one variable are associated with low values of the other, and high values of one variable are associated with high values of the other. Thus we might expect positive correlation between heights and weights of subjects. Negative correlation means that low values of one variable are associated with high values of the other, and high values of one variable with low values of the other. Thus we might expect negative correlation between the heights of a high jump bar and the number of athletes who can clear that bar.

We illustrate the calculation of the correlation coefficient using the data previously met in Section 11.2.

Worked example

11.2 Suppose we consider the coursework marks and examination marks for ten students on the same course at university, as shown in Table 11.8.

Table 11.8.
Coursework and examination marks for ten students

Student	Coursework mark (%)	Examination mark (%)
A	52	50
B	60	62
C	75	70
D	40	50
E	50	55
F	48	50
G	55	58
H	70	70
I	60	40
J	80	75

The scatter diagram was seen previously in Figure 11.1 and shows a strong correlation between the two variables. Confirm this correlation by calculating the correlation coefficient.

Solution   The summations are obtained in tabular form:

x	y	x^2	y^2	xy
52	50	2 704	2 500	2 600
60	62	3 600	3 844	3 720
75	70	5 625	4 900	5 250
40	50	1 600	2 500	2 000
50	55	2 500	3 025	2 750
48	50	2 304	2 500	2 400
55	58	3 025	3 364	3 190
70	70	4 900	4 900	4 900
60	40	3 600	1 600	2 400
80	75	6 400	5 625	6 000
590	580	36 258	34 758	35 210

The correlation coefficient is

$$r = \frac{10(35\,210) - (590)(580)}{\sqrt{\{[10(36\,258) - (590)^2][10(34\,758) - (580)^2]\}}}$$

$$= \frac{352\,100 - 342\,200}{\sqrt{\{[362\,580 - 348\,100][347\,580 - 336\,400]\}}}$$

$$= \frac{9\,900}{\sqrt{\{[14\,480][11\,180]\}}}$$

$$= \frac{9\,900}{\sqrt{161\,886\,400}}$$

$$= \frac{9\,900}{12723.46}$$

$$= 0.78$$

This confirms that there is strong positive correlation between the coursework marks and examination marks.

Worked example

11.3   The data for worked example 11.1, given by

Weeks of experience (x)	4	5	7	9	10	11	12	14
Number of rejects (y)	21	22	15	18	14	14	11	13

were illustrated by Figure 11.2. Calculate the correlation coefficient for the data and interpret.

Solution

x	y	x^2	y^2	xy
4	21	16	441	84
5	22	25	484	110
7	15	49	225	105
9	18	81	324	162
10	14	100	196	140
11	14	121	196	154
12	11	144	121	132
14	13	196	169	182
72	128	732	2156	1069

$$r = \frac{8(1069) - (72)(128)}{\sqrt{\{[8(732) - (72)^2][8(2156) - (128)^2]\}}}$$

$$= \frac{-664}{\sqrt{[(672)(864)]}}$$

$$= \frac{-664}{761.98}$$

$$= -0.87$$

This indicates a good negative correlation between the number of rejects and the number of weeks of experience of the eight employees. The scatter diagram confirms that as the number of weeks of experience increases, then the number of rejects decreases.

Self-assessment question 11.3

1. As part of a manpower planning exercise data were collected on units of output and manpower usage per week for a product during the past week. Calculate the correlation coefficient and comment on the strength of correlation between the two variables.

Output	60	6	12	68	54	42	30	24	40	26	20	32
Manhours used	600	252	260	660	504	480	364	240	400	360	360	480

Exercise 11.3

1. Ten car drivers were each asked how many cars they had owned and how long they had held a full licence, with the results shown.

	Person									
	A	B	C	D	E	F	G	H	I	J
Cars	6	2	8	4	1	1	15	7	9	3
Years	10	1	15	23	1	2	4	10	17	4

Calculate the correlation coefficient and comment appropriately.

2. Two friends were comparing their assessment marks at the end of the year. Ben and Bev's marks are shown in Table 11.9:

Table 11.9. Assessment marks for two students

Subject	Ben	Bev
Mathematics	62	84
Computing	59	40
Business	75	72
Study skills	56	56
Social studies	70	50
Accounting	68	64

Calculate the correlation coefficient and interpret it.

3. Coursework and examination marks for eight students are available for a given module and shown in Table 11.10.

Table 11.10. Coursework and examination marks for eight students

Student	Coursework mark (%)	Examination mark (%)
A	70	50
B	63	80
C	74	54
D	65	45
E	59	59
F	57	47
G	65	64
H	56	77

Calculate the correlation coefficient and interpret the resulting value.

Test and assignment exercises 11

1. The number of marriages in the UK can be subdivided into those that are first marriages for both partners and those that are second marriages for both partners. This information is shown in Table 11.11. Represent the data on a scatter diagram and interpret it.

Table 11.11. Marriage data

Year	Number (000) of marriages first marriage for both partners	Number (000) of marriages second marriage for both partners
1961	340	21
1971	369	36
1981	263	61
1991	222	57

2. Over a period of time a publishing house record the sales, y thousand, of 10 similar textbooks, and the amount, £x hundred, spent on advertising each book. The following table shows the data for the 10 books:

x	0.75	3.90	1.65	1.60	4.40	3.05	3.55	2.65	0.45	2.00
y	2.00	5.35	3.00	2.40	5.95	4.50	4.60	3.65	1.30	3.25

Draw a scatter diagram and calculate the correlation coefficient.

3. The heat output of wood is known to vary with the percentage moisture content. Table 11.12 shows, in suitable units, the data obtained from an experiment carried out to assess this variation.

Table 11.12. Heat output and moisture content for wood

Percentage moisture content, x	Heat output, y
50	5.5
8	7.4
34	6.2
22	6.8
45	5.5
15	7.1
74	4.4
82	3.9
60	4.9
30	6.3

Plot a scatter diagram. Calculate the correlation coefficient and carefully interpret its value.

4. The wind speed and race time were recorded for ten of the races for the British 110 m hurdler Colin Jackson in 1990.

Wind speed (m/s)	−2.9	−1.6	−0.8	−0.4	−0.1	0.2	0.5	1.0	1.1	2.2
Time	13.53	13.39	13.63	13.25	13.63	13.18	13.38	13.11	13.20	13.22

A negative wind speed indicates a head wind, while a positive wind speed is at the runner's back.
(a) Draw a scatter diagram.
(b) Calculate the correlation coefficient.
(c) Interpret your answers to (a) and (b).

5. The car age and costs of maintenance and repairs are available for seven cars, and shown in Table 11.13. Calculate the correlation coefficient and interpret its value.

Table 11.13. Age and costs for seven cars

Age (years)	Annual costs (£)
2	240
3	275
4	350
5	325
6	400
5	340
3	290

12 Probability

Objectives	After reading this chapter, you should be able:
	• to describe the ideas of uncertainty and probability
	• to obtain calculated probabilities using the theoretical and experimental approaches
	• to use the multiplication and addition rules of probability in their correct context

12.1 Introduction

In life we often talk about uncertainty in somewhat vague terms. The following might be said in general conversation:

> She will probably be late

> It is quite likely that my team will win

> The chances are the marriage won't last

Such vagueness can be acceptable in everyday conversation but there may be situations when it is important to define probability more precisely. Such situations might include insurance, genetics (the study of how human characteristics can be passed on from generation to generation), pure science, gambling or a variety of other business decision-making processes. For example, if a company is considering a substantial investment in a new venture, management will want to be satisfied that it will be profitable. To decide whether to invest or not, management will need to identify the relevant probabilities or risks involving profitability.

Probability allows us to try to measure uncertainty. A probability is basically a number which measures the chance of an event happening.

12.2   Approaches to probability

Probability is measured on a scale between 0 and 1 (or, sometimes, 0% to 100%). A probability of 0 means absolute impossibility (for example, the probability a person lives for over one thousand years, or the probability of getting a 7 when we throw a standard die containing the numbers 1, 2, 3, 4, 5, 6) and a probability of 1 means absolute certainty (for example, the probability a person lives for less than one thousand years, the probability of obtaining a number less than seven when we throw a standard die). All other probabilities have a value between 0 and 1 which we write as a fraction, or a decimal.

Figure 12.1 shows the linkage between the vague probability statements and the more specific numerical values.

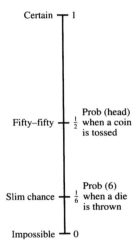

Figure 12.1.
Probability scale

We have now defined what probabilities are, but how can we calculate probabilities in practice? There are two approaches that we will discuss in this text. The theoretical approach assumes that all the outcomes to an event are equally likely, while the experimental approach uses the idea of repeated experimentation to estimate probabilities.

Theoretical approach

This approach assumes that each possible outcome is equally likely to occur.

KEY POINT

The probability that a particular event will happen is

$$= \frac{\text{the number of ways the particular event can happen}}{\text{the total number of outcomes to the experiment}}$$

Problems involving dice are tackled by assuming that, when a die is rolled, each of the six faces is equally likely to be uppermost when it lands. So

$$\text{Prob (1)} = \text{Prob (2)} = \text{Prob (3)} = \text{Prob (4)} = \text{Prob (5)} = \text{Prob (6)} = \tfrac{1}{6}$$

Since 3 out of the 6 equally likely outcomes are even numbers, we would expect an even number $\tfrac{3}{6}$ of the time, so

$$\text{Prob (even number)} = \tfrac{3}{6} = \tfrac{1}{2}$$

Worked example

12.1 A standard pack of 52 playing cards consists of four suits – hearts, clubs, diamonds and spades. Each suit contains 13 cards – an ace, cards numbered from 2 to 10, a jack, a queen and a king. A pack of 52 cards is shuffled thoroughly and the top card is turned up. Write down the probability that this card is

(a) the queen of spades,
(b) a jack,
(c) a plain card (numbered from 2 to 10),
(d) a diamond.

Solution (a) There are 52 cards, but only one of which is the queen of spades, hence

$$\text{Prob (queen of spades)} = \tfrac{1}{52}$$

(b) There are four jacks in the 52 cards, so

$$\text{Prob (jack)} = \tfrac{4}{52} = \tfrac{1}{13}$$

(c) There are $4 \times 9 = 36$ plain cards (from 2 to 10) out of 52, giving

$$\text{Prob (numbered 2 to 10)} = \tfrac{36}{52} = \tfrac{9}{13}$$

(d) There are 13 diamonds, so

$$\text{Prob (diamond)} = \tfrac{13}{52} = \tfrac{1}{4}$$

Note that this answer would have been obtained if it had been identified that there are only four suits, all equally likely, and only one of them is 'diamonds'.

Worked example

12.2 In the UK National Lottery a player chooses six different numbers between 1 and 49 inclusive. At each draw six numbers are generated randomly, and a player wins a share of the jackpot if all the chosen numbers match the six generated numbers. There are 13 983 816 different selections of six different numbers between 1 and 49 and, since the six numbers are drawn at random, the selections are all equally likely to occur.

Find the probability that a player will win a share of the jackpot if
(a) one selection is made,
(b) five different selections are made,
(c) 100 different selections are made.

Solution   (a)   Prob (win) $= \frac{1}{13\,983\,816} = 0.000\,000\,07$, in decimal form

(b)   Prob (win) $= \frac{5}{13\,983\,816} = 0.000\,000\,36$

(c)   Prob (win) $= \frac{100}{13\,983\,816} = 0.000\,007\,15$

Worked example

12.3   Two persons, A and B, unknown to each other previously, meet in a bar. What is the probability that they have the same birthday?

Wow, our birthdays are 59 days apart. I wonder what the chances of that are.

Solution   Whenever A's birthday, there are, ignoring leap years, 365 days on which B might have been born, only one of which coincides with A's.

$$\text{Required probability} = \tfrac{1}{365}$$

Experimental approach

When we are unable to work out the probability using the theoretical approach, then we may use experimental data to estimate the required probability. For example, we can estimate the probability that it will snow in London this Christmas by finding out how many times snow has fallen in London on Christmas Day in the last hundred years, and use the proportion of years on which snow has fallen as an estimate of the probability that it will snow in London on Christmas Day this year.

The experimental approach uses the long-term proportion of identical repeated trials to form an estimate of the required probability.

This approach would be used by insurance companies to estimate the chances that a potential customer will make a claim using information about the claims of similar policy holders. This will influence the insurance companies regarding whether or not to issue a policy and the premium to charge.

Worked example

12.4 Five hundred components have previously been selected randomly from a production line, of which 25 were observed to be below standard. Estimate the probability that a component taken off this production line will be below standard.

Solution Prob (below standard) $= \frac{25}{500} = \frac{1}{20}$

Worked example

12.5 A box of 200 drawing pins were dropped onto a table. Eighty of the drawing pins landed point up, and the remainder landed point down. Estimate the probability that a further drawing pin that is dropped will land point up.

Solution Prob (point up) $= \frac{80}{200} = \frac{2}{5}$

Worked example

12.6 Table 12.1 contains information about the punctuality of the delivery of 400 orders made by three different suppliers.

Table 12.1.
Delivery times for three suppliers

Supplier	Early	On time	Late	Total
Adams	40	20	10	70
Bright	20	120	60	200
Chambers	0	30	100	130
Total	60	170	170	400

(a) Calculate the probability that a delivery chosen at random is early.

(b) Calculate the probability that a delivery chosen at random is a delivery from Bright.

(c) Calculate the probability that a delivery chosen at random is both from Adams and late.

(d) Calculate the probability that a delivery chosen at random is late, given that it came from Adams.

(e) Explain the term 'chosen at random' in the context of the example.

Solution (a) Altogether there are 400 deliveries and 60 of these are early.

$$\text{Prob (early)} = \frac{60}{400} = \frac{3}{20} = 0.15$$

(b) Out of the 400 deliveries, Bright is responsible for delivering 200.

$$\text{Prob (Bright)} = \frac{200}{400} = \frac{1}{2} = 0.5$$

(c) Out of the 400 deliveries, 10 are late deliveries from Adams.

$$\text{Prob (Adams and late)} = \frac{10}{400} = \frac{1}{40} = 0.025$$

(d) Altogether Adams delivers on 70 occasions, of which 10 are late.

$$\text{Prob (late given Adams)} = \frac{10}{70} = \frac{1}{7} = 0.14$$

This is an example of conditional probability, the condition being that the delivery was one by Adams, rather than all deliveries.

(e) Random means purely by chance in an unbiased way. For example, if we select a delivery at random, then all 400 deliveries are equally likely to be chosen.

Self-assessment questions 12.2

1. It is known that there are five winning raffle tickets and that 200 have been sold. What is the chance of winning with one raffle ticket

(a) first prize,
(b) any prize?

2. One-hundred ball-bearings are checked in a factory and 25 are found to be defective. If one is picked at random, what is the chance that it will be

(a) defective,
(b) good?

3. What range of values can a probability take? If a probability is close to zero, what does that mean?

Exercise 12.2

1. A survey of 200 households gave the number of children under 18 living in each household. This information is shown in Table 12.2

Table 12.2. Distribution of children in 200 households

Number of children	Frequency
None	80
One	50
Two	40
Three	20
Four or more	10

Obtain the probability that when a household is chosen at random
(a) it has no children,
(b) it has three or more children,
(c) it has at least one child.

2. A bag contains 40 plastic clothes pegs identical in all but colour; 20 of these are red, 12 are white, and the rest are yellow. Determine the probability that a peg chosen at random from the bag is
(a) yellow,
(b) not white.

3. In a group of 20 students, 7 wear glasses and 4 are left-handed. Two of the students are both left-handed and wear glasses. Draw up a table showing the numbers that wear and do not wear glasses together with those that are left- and right-handed.

A student is chosen at random from the class. Calculate the probability that the student
(a) is right-handed.
(b) does not wear glasses and is right-handed.

4. A product passes through a production line of four consecutive processes before it is completed. In each process, some units are rejected and scrapped. Data for a particular month are shown in Table 12.3.

Table 12.3. Rejection rates on the production line

	Units
Input to production line	5000
Rejected after process 1	600
Rejected after process 2	400
Rejected after process 3	250
Rejected after process 4	100

(a) What is the probability that a unit, once started, will become a completed unit of the finished product?
(b) What is the probability that a unit will enter process 3?

12.3   Probability rules

In the previous section we dealt with simple probability problems. However, for more complex problems, we need to identify certain probability rules. It is useful to formalize three basic probability rules.

'Not' rule

This rule is relevant when we need to find the probability of an event not happening.

> Prob (event does not happen) $= 1-$ Prob (event does happen)

Worked example

12.7 A bag contains five blue balls, three red balls, and two black balls. A ball is drawn at random from the bag, calculate the probability that it will not be black.

Solution   Prob (black) $= \frac{2}{10}$

Therefore the probability that it is not black is

Prob (not black) $= 1 - \frac{2}{10} = \frac{8}{10} = \frac{4}{5}$

'Or' rule

Two events are mutually exclusive if only one can happen. For example, if a coin is tossed and a head appears uppermost then a tail cannot also appear uppermost. Hence a head and a tail are mutually exclusive.

> Providing events A and B are mutually exclusive then
> Prob (A or B) = Prob (A) + Prob (B)

Worked example

12.8 If one card is drawn randomly from a normal pack of 52 playing cards, what is the probability of getting an ace or a king?

Solution As ace and king are mutually exclusive events

Prob (ace or king) = Prob (ace) + Prob (king)

$$= \frac{4}{52} + \frac{4}{52} = \frac{8}{52}$$

Note that ace and spade would not be mutually exclusive events and so the above 'or' rule does not apply.

Worked example

12.9 Suppose a supermarket manager is interested in studying the sales of a particular item. The record of the number of items sold in the last 100 sales days is shown in Table 12.4.

Table 12.4.
Units sold in a
supermarket

Number of items sold	Days
1	4
2	6
3	25
4	35
5	19
6	11
Total	100

(a) Estimate the probability the supermarket sells at least two items on a given day.

(b) Estimate the probability the supermarket sells no more than four items on a specified day.

Solution (a) Prob (at least 2) = Prob (2 or 3 or 4 or 5 or 6)
$$= \text{Prob (2)} + \text{Prob (3)} + \text{Prob (4)}$$
$$+ \text{Prob (5)} + \text{Prob (6)}$$
$$= 0.06 + 0.25 + 0.35 + 0.19 + 0.11$$
$$= 0.96$$

Note that we might have attempted this question by using

Prob (at least 2) = 1 − Prob (1) = 1 − 0.04 = 0.96

(b) Prob (no more than 4) = Prob (1 or 2 or 3 or 4)
$$= \text{Prob (1)} + \text{Prob (2)} + \text{Prob (3)} + \text{Prob (4)}$$
$$= 0.04 + 0.06 + 0.25 + 0.35$$
$$= 0.70$$

'And' rule

Two events are said to be independent if the occurrence of one event does not influence the occurrence of the second event.

For example, the score obtained on rolling a die has no influence on the score obtained on rolling the same die a second time or on rolling a second die, so these are independent events. Also, when looking at the births of boys and girls it is assumed that whether a baby is a boy or a girl is unaffected by whether babies born earlier were girls or boys.

KEY POINT

If two events, A and B, are independent then
$$\text{Prob (A and B)} = \text{Prob (A)} \times \text{Prob (B)}$$

Worked example

12.10 Find the probability that, when two coins are tossed, they both land heads up.

Solution Whether or not one coin lands heads up is clearly not influenced by whether the other lands heads up, so the two events are independent, with each coin having the probability $\frac{1}{2}$ of landing heads up.

$$\text{Prob (both coins heads up)} = \text{Prob (first coin heads up)}$$
$$\times \text{Prob (second coin heads up)}$$

$$= \tfrac{1}{2} \times \tfrac{1}{2} = \tfrac{1}{4}$$

Worked example

12.11 Use the 'and' rule to estimate the probability that the first three children in a family are all girls.

Solution Assuming that girls and boys are equally likely, then at each birth

$$\text{Prob (boy)} = \text{Prob (girl)} = 0.5$$

Then

Prob (first three babies are girls)
= Prob (first is a girl *and* second is a girl *and* third is a girl)

We now assume that the births are independent of each other, so

Prob (first three babies are girls)

= Prob (first is a girl) × Prob (second is a girl) × Prob (third is a girl)

$= 0.5 \times 0.5 \times 0.5 = 0.125$ (or $\frac{1}{8}$)

Worked example

12.12   On the basis of life-expectancy tables, the probability that a husband will be alive in ten years' time is 0.8, and the probability that his wife will be alive in ten years' time is 0.9. Find the probability that, in ten years time,
(a) both will be alive,
(b) neither will be alive.

Solution   (a) Prob (both alive) = Prob (husband alive *and* wife alive)

If we assume that the two events are independent, then this equals

Prob (husband alive) Prob (wife alive) $= 0.8 \times 0.9 = 0.72$

(b) Prob (neither alive) = Prob (husband dead *and* wife dead)

$= 0.2 \times 0.1 = 0.02$

There are a number of situations in real life in which events are not mutually exclusive or independent, and for which the above probability rules will need to be adapted. These are covered in more advanced textbooks.

Self-assessment questions 12.3

1. Explain the terms (a) mutually exclusive events, (b) independent events.

2. (a) A fair die is rolled. State the probability of getting a five or more.
   (b) Two fair dice are rolled. Obtain the probability that both score sixes.

Exercise 12.3

1. A dealer of shirts has white, green, yellow, blue and pink shirts for sale. The sales indicate that when a person buys a shirt, the probability that he chooses a white, a green, a yellow, a blue, and a pink shirt are 0.5, 0.2, 0.1, 0.15, 0.05, respectively. Find the probability that he will choose:
   (a) either a blue or green shirt,
   (b) not a white shirt,
   (c) neither a pink nor a yellow shirt.

2. A bag contains 20 white balls, 10 red balls and 10 blue balls. A ball is drawn at random, replaced, and a second ball is drawn at random. Find the probability that:
   (a) both balls are red,
   (b) the first ball is white and the second ball is blue.

3. A butcher knows that the sales demand for fresh chickens is four, five or six chickens per day. The probability for demand is:

Demand per day	Probability
Four chickens	0.5
Five chickens	0.3
Six chickens	0.2

Demand on one day is independent of demand on any other day.

(a) What is the probability that on any day there are more than four chickens sold.

(b) What is the probability the demand will be five chickens on each of two consecutive days?

(c) What is the probability that demand will be four chickens on each of day 1 and day 2, and six chickens on day 3?

4. A market researcher has established that one in four cyclists use Long-glow batteries in their cycle lamps. She then selects two cyclists at random to interview about their batteries. What is the probability that neither of the cyclists use Long-glow batteries.

5. (a) One card is drawn at random from a full pack of 52 cards. What is the probability it is not a spade.

   (b) The first card is replaced and a second card is drawn at random. What is the probability that both cards are spades?

Test and assignment exercises 12

1. A record of the sex of customers at a shoe shop on a particular morning and whether they bought anything or not is given in Table 12.5.

Table 12.5. Buying frequency for the two sexes

	Bought	Did not buy
Male	10	5
Female	13	8

Calculate the probability that, on a particular morning, a customer chosen at random

(a) was female,

(b) did not buy anything,

(c) was male and bought something.

2. John has five unbiased coins in his pocket. He takes each coin in turn and tosses it.

   (a) What is the probability of the first coin landing heads?

   (b) The first four coins land heads. What is the probability of the fifth coin landing heads?

   (c) The fifth coin lands heads. What is the probability of tossing any five unbiased coins and getting five heads?

3. At a darts club presentation evening you are invited to draw a raffle ticket out of a hat in exchange for £1. To win a prize, the number on the ticket had to end in 0 or 5. Suppose there were 500 tickets in the hat numbered from 1 to 500, what was the probability that you would win a prize?

4. A simple type of fruit machine is devised for introducing the ideas of probability to students. The machine has two windows and at each window is shown a picture of either a cherry or an orange. This is done by a random spinning of wheels in the machine, each wheel having five pictures. On wheel 1 there are two cherries and three oranges and on wheel 2 there is one cherry and four oranges.
   (a) State the probability that a cherry appears at window 1 after a random spinning of this wheel.
   (b) Find the probability of obtaining two cherries at the two windows.
   (c) A person wins if the same fruit appears on both wheels. Determine the probability of a person winning on one go of the fruit machine.

5. A shoe shop keeps a record of the colour and size of men's shoes sold throughout the year. The data for the month of August are given in Table 12.6.

Table 12.6. Colour and size of men's shoes

Size	Black	Brown	Grey	Other
6	1	2	0	0
7	4	3	1	1
8	7	5	2	2
9	10	15	4	1
10	9	12	5	2
11	5	8	1	0

Calculate the probability that a customer who bought a pair of men's shoes from the shop in August bought a pair that was
   (a) black,
   (b) brown size 10,
   (c) a colour other than black, brown or grey,
   (d) size 9 given that it was brown,
   (e) size 10 or over,
   (f) size 8 or 9 but not grey,
   (g) grey size 8 or black size 10.

6. An item produced by a company is subject to two possible types of defect, A (an electrical fault) or B (a non-electrical defect). The probability that the item has defect A is $\frac{1}{8}$ and the probability it has defect B is $\frac{1}{6}$, independent of whether it has defect A.
   (a) Find the probability that an item has both A and B defects.
   (b) Find the probability that an item has no defect.

7. An insurance company divides its clients into two age groups, 'under 25' and '25 and over'. In a particular year 120 of the 500 clients were aged under 25. After one year, 150 of the clients had made a claim on their insurance, of whom 50 were under 25. Find the probability that a randomly selected client
   (a) is under 25,
   (b) has made a claim,
   (c) is under 25 and not made a claim,
   (d) has made a claim given that he is under 25.

8. An item is made in three stages. At the first stage it is formed on one of four machines A, B, C, D with equal probability. At the second stage it is trimmed on one of three machines E, F and G, with equal probability. Finally, it is polished on one of two polishers H and I, and is twice as likely to be polished on the former, as this machine works twice as quickly as the other. What is the probability that an item is:
   (a) polished on H?
   (b) trimmed on either F or G?
   (c) formed on either A or B, trimmed on F and polished on H?

9. The probability that A can solve a problem is $\frac{2}{3}$ and the probability that B can solve it is $\frac{4}{5}$. If both try, what is the probability that the problem is solved?

10. An insurance agent finds that on following up an enquiry the probability of making a sale is 0.4. If, on a particular day, the agent has two independent enquiries, what is the probability that he will sell
(a) insurance to both enquirers,
(b) no policies?

Section D
Algebra

13 Simple algebra

Objectives

After reading this chapter, you should be able:

- to use and read simple algebraic notation
- to substitute values into simple algebraic expressions
- to simplify algebraic expressions

13.1   Introduction

Algebra is concerned with the manipulation of symbols. These symbols are chosen to represent either variables or constants. To represent speed (a variable) we may choose the symbol s, and to represent a fixed interest rate (a constant) we may choose the symbol R. In this chapter we introduce some of the basic ideas of algebra which are essential for more advanced study of mathematics or statistics.

13.2   Terminology

Whenever letters or symbols are used to represent numbers, or sets of numbers, then the resulting expressions are called *algebraic* expressions. For example,

$$C = \tfrac{5}{9}(F - 32)$$

where C is degrees Celsius and F is degrees Fahrenheit, or

$$A = \pi r^2$$

where A is the area of a circle with radius r, are both examples of algebraic expressions. Algebra enables answers to be found to problems by simple operations with letters rather than by repeated arithmetic with numbers.

An algebraic expression can be considered to be a set of letters and numbers combined by the arithmetic operators $+$, $-$, $\times$, $\div$. For example, in the manufacture of a certain product, the profit in pounds, obtained by making and selling x products may be represented by

$$2x - 50$$

This is an algebraic expression with two terms, $2x$ and 50. The *terms* of an algebraic expression are those parts of it that are connected by $+$ or $-$ signs. Further definitions necessary for the understanding of algebra are

1. *Variables:* letters used to represent different numbers, e.g. in $2x - 50$ there is just one variable, x.
2. *Coefficient:* a number placed before and thus multiplying a letter or group of letters, e.g. in $2x - 50$ the coefficient of x is 2.
3. *Constant term:* a term with no variables, e.g. in $2x - 50$, a constant term is -50.

Occasionally you will meet a constant that is represented by a symbol. A very common example is π. Here π is a constant, which is approximately equal to 3.142. In the following expression x is a variable, a is a constant and b is a coefficient:

$$a + bx$$

KEY POINT

In the expression $5x - 7$ the components are:

variable	x
coefficient	5
constant	-7

It is particularly important that the meaning of algebraic expressions is clearly understood. Some examples of algebraic expressions, together with their meanings, are given in Table 13.1. Note that in algebra the multiplication sign, $\times$, is usually omitted to save confusion with the letter x.

Table 13.1.
Algebraic terminology

Algebraic expression	Interpretation
x	$1(x)$
$2x$	$2(x)$ or $x + x$
ax	$a(x)$
$\frac{1}{2}x$	$\frac{1}{2}(x)$ or $x/2$
$-x$	$-1(x)$ or $-1x$
$2ax$	$(2a)x$ or $ax + ax$
x^2	$x(x)$
$2x^2$	$2x(x)$ or $x^2 + x^2$
$x^{\frac{1}{2}}$	$\sqrt{x}$
$(2x)^2$	$(2x)(2x)$ or $(2^2)x^2$ or $4x^2$
$2ax^2$	$2a(x^2)$ or $ax^2 + ax^2$
$2x + 3y$	$2(x) + 3(y)$

Substitution is the replacing of letters in an algebraic expression by given values to obtain a numerical value for that expression. The algebraic expression that determines the temperature in degrees Celsius from the temperature in degrees Fahrenheit was stated at the beginning of this unit. What is the equivalent temperature in degrees Celsius when the temperature is 59 °F? Replace F in the expression

$$C = \tfrac{5}{9}(F - 32)$$

with the value 59. Then

$$C = \tfrac{5}{9}(59 - 32)$$
$$= \tfrac{5}{9}(27) = 15$$

Therefore 59 °F is equivalent to 15 °C. In a similar way, the equivalent temperature to 70 °F is

$$C = \tfrac{5}{9}(70 - 32)$$
$$= \tfrac{5}{9}(38) = 21.1$$

that is, 21.1 °C.

Worked example

13.1 The area of a trapezium is given by the algebraic expression

$$\tfrac{1}{2}(x + y)h$$

where x and y are the lengths of its two parallel sides and h is the perpendicular distance between these two sides. Find the area of the trapezium given in Figure 13.1.

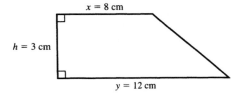

Figure 13.1.
Trapezium

Solution   In this example $x = 8$, $y = 12$, $h = 3$, so the area of the trapezium is

$$\tfrac{1}{2}(8 + 12)3 = \tfrac{1}{2}(20)3 = 30\,\text{cm}^2$$

Worked example

13.2   If $x = 3$, $y = 2$ and $z = -1$, find the values of

(a) $2xy + z$
(b) $3x^2 + 2yz$
(c) $4x - \tfrac{1}{2}y + (2z)^2$

Solution   (a) $2xy + z = 2(3)(2) + (-1)$
$\qquad\qquad\quad = 12 - 1 = 11$

(b) $3x^2 + 2yz = 3(3^2) + 2(2)(-1)$
$\qquad\qquad\quad = 27 - 4 = 23$

(c) $4x - \tfrac{1}{2}y + (2z)^2 = 4(3) - \tfrac{1}{2}(2) + (2(-1))^2$
$\qquad\qquad\qquad\qquad = 12 - 1 + (-2)^2$
$\qquad\qquad\qquad\qquad = 12 - 1 + 4 = 15$

Do not forget that $(-2)^2 = (-2)(-2) = 4$ as multiplication involving two negative numbers gives a positive answer.

KEY POINT

> The square of a negative number is always positive: that is,
> $$(-x)(-x) = +x^2$$

Self-assessment questions 13.2

1. Suggest suitable symbols for
   (a) Temperature   (b) Time   (c) Price   (d) Profit

2. If $x = 3$ evaluate $24 - 5x$

3. Evaluate $6D + T^2$ if $D = 2$ and $T = -3$.

Exercise 13.2

1. If $x = 5$, $y = -3$ and $z = 1$ evaluate
   (a) xyz (b) zxy
   (c) $x + y + z$ (d) $x(y + z)$
   (e) $xy + z$ (f) $xy + (z)$
   (g) $(x + z)/y$

2. If $w = 2$, $t = 3$ and $s = -4$ evaluate
   (a) $3w^2$ (b) $3(w^2)$
   (c) $(3w)^2$ (d) $t^2 + w^2$
   (e) $(t + w)^2$ (f) $st^2 - tw^2$

13.3   Simplification

Suppose we wish to evaluate

$$5x - 3x$$

when $x = 7$. It could be done by substituting $x = 7$ into this expression:

$$5(7) - 3(7) = 35 - 21 = 14$$

However, it is more straightforward to simplify the original expression to get

$$5x - 3x = (5 - 3)x$$
$$= 2x$$

and then substitute $x = 7$ into this expression,

$$2(7) = 14$$

The aim of simplification is to convert an algebraic expression into one that is shorter and easier to handle. The terms $5x$ and $-3x$ are called like terms because the only difference is their coefficients. For example,

$$3y, 5y \text{ and } -6y$$

are like terms. Only like terms can be added or subtracted as a single term, the process being called 'collecting like terms together'. For example,

$$8a + 6a - 3a + a = 12a$$

KEY POINT

$3x$, $3(x)$, $x3$, $x(3)$ are all the same and usually written $3x$.

However, $3x$ and $2y$ are not like terms and so $3x - 2y$ cannot be simplified in this way. However, the order in which the letters are written in like terms is not important, for example,

$$4xyz - 2yzx + 7zyx = 9xyz$$

as xyz, yzx and zyx are also like terms. A useful tip when dealing with more complicated terms is always to write the letters in alphabetical order.

KEY POINT

xy, yx, $x(y)$, $y(x)$, $(y)x$ and $(x)y$ are the same and are usually written as just xy.

When multiplying or dividing algebraic expressions containing like letters, the basic rules of powers as described in Chapter 4 apply.

Therefore

$$(9x^2yz)(2xy^3z) \div 3xyz = 18x^3y^4z^2 \div 3xyz$$
$$= \frac{18x^3y^4z^2}{3xyz}$$
$$= 6x^2y^3z$$

If you had problems following this last example, go back to Chapter 4 and remind yourself of the rules for powers.

Worked example

13.3 Simplify

(a) $4x^2y - 3x^2y + x^2y$
(b) $4ab(-2bc)$
(c) $6mn^2 \div 2m^2n$

Solution   (a) $4x^2y - 3x^2y + x^2y = (4 - 3 + 1)x^2y = 2x^2y$

(b) $4ab(-2bc) = -8ab^2c$

(c) $6mn^2 \div 2m^2n = \dfrac{6mn^2}{2m^2n} = \dfrac{3n}{m}$

Brackets, or parentheses, are often used in algebraic expressions as they are when performing numerical calculations. They may be removed by multiplying the term outside the bracket by each of the terms inside the bracket. For example

$$3(x + y) = 3x + 3y$$

In particular, care must be taken with the sign of each term when removing brackets; a negative sign outside a bracket changes all the signs inside the bracket. For example

$$-3(x - y + z) = -3x + 3y - 3z$$

KEY POINT | $a(x + y)$ and $(x + y)a$ are the same and $a(x + y) = ax + ay$

Worked example

13.4 Simplify

(a) $5x^2 - 2x(4 - 3x) - 3x$
(b) $3xy - y(2x + 4)$

Solution   (a) $5x^2 - 2x(4 - 3x) - 3x = 5x^2 - 8x + 6x^2 - 3x$
$$= 11x^2 - 11x$$

which could be written $11x(x - 1)$.

Do not be overambitious when simplifying algebraic expressions involving brackets. First remove the brackets and then collect like terms together. Do not attempt to carry out both processes in one step.

(b) $3xy - y(2x + 4) = 3xy - 2xy - 4y$
$$= xy - 4y$$
$$= (x - 4)y$$

KEY POINT

> The product of two brackets is obtained by multiplying each term in the first bracket by each term in the second. For example,
> $$(a + b)(x + y) = ax + bx + ay + by$$

Worked example

13.5   Expand

(a) $(x - 1)(x + 2)$
(b) $(a - b)(a + b)$
(c) $(x + y + z)^2$

Solution   (a) $(x - 1)(x + 2) = x^2 - x + 2x - 2$
$$= x^2 + x - 2$$

(b) $(a - b)(a + b) = a^2 - ba + ab - b^2$
$$= a^2 - b^2$$

(c) $(x + y + z)^2 = (x + y + z)(x + y + z)$
$$= x^2 + yx + zx + yx + y^2 + yz + zx + zy + z^2$$
$$= x^2 + 2xy + 2xz + y^2 + 2yz + z^2$$

Algebra is a useful tool for translating verbal statements into mathematical form. For example, if one kilo of apples costs 48p, an algebraic expression for the cost of x kilos is (in pence) $48x$.

Similarly, if one kilo of carrots costs 35 pence and one cauliflower costs 18 pence, then the total cost of p kilos of carrots and q cauliflowers is (in pence)

$$35p + 18q$$

When forming an algebraic expression from a description it is a useful idea to substitute numerical values into the expression to check that it is correct.

Worked example

13.6   Under normal driving conditions, an approximation for the minimum braking distance (in metres) of a car is obtained by dividing the square of the speed (in km/h) by 200.

(a) Express a relationship between the minimum braking distance, d metres, and the speed, v km/h.

(b) Find the minimum braking distances for speeds of 40 km/h and 80 km/h.

Solution   (a) $d = \frac{1}{200} v^2$

(b) When $v = 40$,

$$d = \frac{1}{200}(40)40 = 8 \text{ metres}$$

and when $v = 80$,

$$d = (80)80 = 32 \text{ metres}$$

Note that doubling the speed from 40 km/h to 80 km/h results in a braking distance that is four times the original braking distance.

Worked example

13.7 Mr Davies keeps a very careful record of his motoring expenses. The fixed costs of possessing his car (which includes road tax and insurance) are £350 per year. His variable costs (which includes petrol, oil and maintenance) are 12 pence per mile. There is public transport from his home to all the places he normally travels to by car, the fare amounting to 18 pence per mile.

(a) Using the symbol x to represent the number of miles travelled by Mr Davies in a year, give an expression for his annual travelling costs
   (i) if he only travels by car,
   (ii) if he sells his car and travels by public transport.

(b) Decide which method of transport costs less if he travels
   (i) 5000 miles,
   (ii) 10 000 miles.

Solution   (a) (i) If he travels by car,
Cost (in £) $= 350 + 0.12x$
Be careful with units.

(ii) Cost (in £) $= 0.18x$
if public transport is used.

(b) (i) If $x = 5000$
Cost using car $= 350 + 0.12(5000) = £950$
Cost using public transport $= 0.18(5000) = £900$
If the distance travelled is 5000 miles, public transport is cheaper.

(ii) If $x = 10\,000$
Cost using car $= 350 + 0.12(10\,000) = £1550$
Cost using public transport $= 0.18(10\,000) = £1800$
If the distance travelled is 10 000 miles, car travel is cheaper. (No account has been taken of the money obtained from the sale of his car.)

Self-assessment questions 13.3

1. Is it true that $5x^2y = x(5y)x$?

2. Is it true that $4y^3x^2 = y(2y)^2y^2$?

3. Is it true that $4(2x - 5y) = -20y + 8x$?

Exercise 13.3

1. Simplify the following:
(a) $3xy + yx8$
(b) $5yx - 3xy + x5y$
(c) $6x + 2y - 4x - 2y$

2. Simplify the following:
(a) $2(x - 3y) + 6y$
(b) $9x^2y \div 3xy$
(c) $(4xyz)/(2xy)$

3. Simplify the following:
(a) $4pq + 6p^2q \div p$
(b) $4(a + 3b) + 6(2b - a)$
(c) $(v + w)^2 \div (w + v)$

4. A quarter's telephone bill involves a rental charge of £25.00 plus a further charge of 5.5p per metered unit.
(a) Use the symbol x to represent the number of metered units and hence give an expression showing the total quarterly cost of using x units.
(b) Find the quarterly cost if
(i) 400 units
(ii) 1000 units
are used during the quarter.
(c) An independent communications company offers an alternative pricing policy – a quarterly rental charge of £40.00 plus a charge per unit of 3p. For each of the two values in (b) work out whether the new offer is worth while.

Test and assignment exercises 13

1. The volume of a cylinder is given by

   $$v = \pi r^2 h$$

   where r is the radius of the cylinder and h is the height of the cylinder. Calculate the volume of a cylindrical tank with height 3 metres and radius $\frac{1}{2}$ metre.

2. The relationship between °F and °C is given by

   $$C = \frac{5}{9}(F - 32)$$

   If $F = 45°$ determine the equivalent temperature in °C.

3. If $x = 5$, $y = 2$, find values of
   (a) $5x^2 + 2xy - y^2$
   (b) $3x^2y$
   (c) $5x + 4y - 8$

4. Simplify
   (a) $8x - 5x + 3x$
   (b) $2x^2y - 5x^2y + yx^2$
   (c) $5x + 8y$

5. Simplify
   (a) $4xy(2x^2y)$
   (b) $8ab^3 \div 2a^3b$
   (c) $5p(8q) \div 4pq$

6. One litre of petrol costs 69.5p and one can of oil costs £1.30. Write down an expression for the cost of x litres of petrol and y cans of oil.
   Determine this cost when $x = 6$ and $y = 2$.

7. Expand
   (a) $(x + y)^2$
   (b) $5x - 2(x - 2) - 5$
   (c) $(a + b - ab)(a - b)$

8. A small manufacturer makes electrical components. He has fixed costs of £800 per month (including rent and rates) and variable costs of 60 pence per component. At the end of each month a wholesaler buys as many components as the company can produce, paying £1 per component. If the manufacturer makes x components in a month, write algebraic expressions for

   (a) the monthly costs,
   (b) the monthly revenue.

   Determine the monthly costs and monthly revenue if he manufactures 1000, 2000 and 3000 components Comment on your answer.

9. Simplify $2(x^2 - x + 3) - (x^2 + 2x - 4)$.

10. A rocket accelerates upwards such that h, its height in metres, after t seconds is given approximately by the formula

    $$h = 12t^2$$

    Calculate its height after 5 seconds and after 20 seconds. What is the average velocity during this time period? Obtain an algebraic expression that determines the average velocity, between time a seconds and b seconds.

14 Linear equations

<table>
<tr><td>Objectives</td><td>After reading this chapter, you should be able:

• to use algebra to formulate linear equations

• to solve linear equations</td></tr>
</table>

14.1   Introduction

An equation is an algebraic statement that two quantities are equal. In Chapter 13 we saw the equation:

$$C = \tfrac{5}{9}(F - 32)$$

Here there are two variables C and F and for a particular value of F we can find the corresponding value for C. If $F = 32$ then $C = 0$. This is an example of a simple linear equation. It is called linear because when its graph is drawn a straight line is produced.

14.2   Solving linear equations

The monthly repayments needed to repay a mortgage of 25 years are shown in Table 14.1. This has already been discussed in Chapter 8, and it is clear from Figure 8.2 that the relationship between the amount of loan and the monthly repayment is linear, i.e. the graph is a straight line.

Table 14.1.
Calendar monthly
repayments at 8.925% p.a.

Amount of loan (£000)	1	5	10	20
Monthly repayment (£)	8.45	42.25	84.50	169.00

Before discussing this further, it is necessary to introduce some symbols. Let L be the size of the loan, measured in thousands of pounds, and R the monthly repayment in pounds, so that we have, from Table 14.1, if $L = 5$, $R = 42.25$.

KEY POINT   The graph of a linear equation is a straight line.

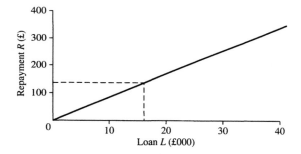

Figure 14.1.
Loan repayments

To find the repayment necessary for a loan of £16 500 Figure 14.1 could be used, as in Chapter 8, but this will only give an approximate answer. A better approach would be to find an equation that will allow R to be calculated from L. Now when $L = 1$, $R = 8.45$ and when L is increased by a multiple of 10 to 10, R is increased by a multiple of 10 to 84.50. Also when L doubles from 10 to 20, so does R from 84.50 to 169.00. These results are the consequences of the linear relationship between L and R, and it is possible to deduce that if $L = 2$, then R will be $2(8.45) = 16.9$, while if $L = 3$, $R = 3(8.45) = 25.35$. The algebraic relationship between L and R, or the linear equation giving R in terms of L, is

$$R = 8.45(L)$$

or

$$R = 8.45L$$

This equation can be used to give accurate values for the monthly repayments. A loan of £16 500 ($L = 16.5$) will need a monthly repayment of $R = 8.45(16.5) = 139.43$, while a loan of £37 950 ($L = 37.95$) will need a monthly repayment of $R = 8.45(37.95) = 320.678$, i.e. £320.68. (This explains the accurate results stated in Chapter 8.)

To manufacture a large batch of electronic components there are two costs incurred, a fixed set-up cost of £500.00 and a variable cost due to labour and materials of £3.00 for each component. Hence if a batch of 100 components is made the total cost is $500 + 3(100) = £800.00$. Let the number of components in a batch be x and let the total cost be C. By calculating C for various values of x the results in Table 14.2 were obtained, and the graph of these results is in Figure 14.2. This graph shows that there is a linear relationship between C and x.

Table 14.2.
Costs for various batch sizes

Batch size, x	10	100	200	300
Cost, C (£)	530	800	1100	1400

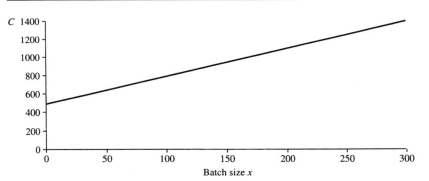

Figure 14.2.
Costs for various batch sizes.

The linear equation connecting C and x is

$$C = 500 + 3x$$

You should use this equation to check the results in Table 14.2; for example, if $x = 200$,

$$C = 500 + 3(200)$$

i.e.

$$C = 500 + 600$$

$\therefore$ $C = 1100$

If the total cost is to be £1010, what should the batch size be? This is equivalent to finding x when $C = 1010$, or solving the equation

$$500 + 3x = 1010$$

The graph can be used to give an approximate answer by reading off the value of x that corresponds to $C = 1010$, but a more accurate method is to use the rules of algebra (Chapter 13). An equation remains true as long as whatever is done to one side is also done to the other. Subtracting 500 from both sides produces:

$$3x = 510$$

Dividing both sides by 3 produces:

$$x = \tfrac{510}{3} = 170$$

Thus a cost of £1010 is incurred if a batch of 170 is produced.

A linear equation such as $8 - 2x = 0$ can be solved by algebra. In this case the solution is $x = 4$.

Worked example

14.1   The two variables x and y are related by the following linear equation:

$$y = 15 - 2x$$

(a) Find y if $x = 4$.
(b) Find x if $y = 5$.
(c) Find x if $y = 0$.

Solution   (a) If $x = 4$, then $y = 15 - 2(4)$
$$= 15 - 8$$
$$= 7$$
$\therefore\ y = 7$ when $x = 4$

(b) If $y = 5$, then $5 = 15 - 2x$
Add $2x$ to both sides: $5 + 2x = 15$
Subtract 5 from both sides: $2x = 10$
Divide both sides by 2: $x = 5$
$\therefore\ x = 5$ when $y = 5$

(c) If $y = 0$, then $0 = 15 - 2x$
$\therefore \qquad\qquad 2x = 15$
$\therefore \qquad\qquad x = \tfrac{15}{2}$
$\therefore \qquad x = 7.5$ when $y = 0$

Both of the previous linear equations have been of the form $y = a + bx$ where a and b are constants. For example, the equation $y = 15 - 2x$ has $a = 15$ and $b = -2$. The equation $y = a + bx$ is typical of linear equations which predict y from x, and Figure 14.3 shows the graphs of three such equations, each of which is discussed in detail.

1. $y = 5 + x$. Here $a = 5$ and $b = 1$. If $x = 0$, $y = 5$, which is the value of a. If $x = 10$, $y = 15$ and so y increases by 10 when x increases by 10. Hence the gradient is $10/10 = 1 = b$.

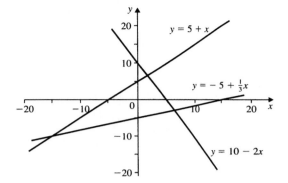

Figure 14.3.
Graph of linear equations

2. $y = -5 + \frac{1}{3}x$. Here $a = -5$ and $b = \frac{1}{3}$. If $x = 0$, $y = -5$, which is the value of a. If $x = 10$,

$$y = -5 + \frac{10}{3} = -5 + 3\frac{1}{3} = -1\frac{2}{3}$$

and so y increases by $3\frac{1}{3}$ when x increases by 10. Hence the gradient is

$$\frac{10}{3} \div 10 = \frac{10}{3}\left(\frac{1}{10}\right) = \frac{1}{3} = b$$

3. $y = 10 - 2x$. Here $a = 10$ and $b = -2$. If $x = 0$, $y = 10$, which is the value of a. If $x = 10$, $y = 10 - 20 = -10$, and so y decreases by 20, hence the gradient is $-20/10 = -2 = b$.

These three examples illustrate the general result that for the graph of the linear equation $y = a + bx$, a is the *intercept* on the y axis (value of y when $x = 0$), and b is the *gradient*.

KEY POINT

> The usual form for a linear equation is
> $$y = a + bx$$
> where a is the intercept and b is the gradient and is represented in Figure 14.4.

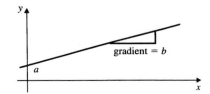

Figure 14.4.
The general straight line

Worked example

14.2   The relationship between x and y is linear. When $x = 0$, $y = 3$ and when $x = 4$, $y = 19$. Find the equation relating y to x.

Solution   Knowing two points that the straight line goes through, the graph could be drawn, since it is a straight line. From this graph the intercept (a) and gradient (b) could be measured, and so the equation $y = a + bx$ determined. In this example, however, both a and b can be found without a graph. Since $y = 3$ when $x = 0$, the intercept is 3, i.e. $a = 3$. Since y increases by $19 - 3 = 16$ when x increases by 4, the gradient is $16/4 = 4$. Thus $b = 4$ and the required equation is:

$$y = 3 + 4x$$

A gardener is planning to grow a mixture of gooseberry and blackcurrant bushes on a plot of land of area $300 \, \mathrm{m}^2$. Each gooseberry bush requires $3 \, \mathrm{m}^2$ while each blackcurrant bush requires $5 \, \mathrm{m}^2$. Let G and B be the numbers of gooseberry and blackcurrant bushes respectively. Then if all the $300 \, \mathrm{m}^2$ is to be used, the following linear equation must hold:

$$3G + 5B = 300$$

If $G = 0$, then $5B = 300$ or $B = 60$.

If $B = 0$, then $3G = 300$ or $G = 100$.

If $B = 30$, then $3G + 5(30) = 300$, i.e. $3G + 150 = 300$.

Subtract 150 from both sides:

$$3G = 150$$

and divide both sides by 3:

$$G = 50$$

I don't like blackcurrants or gooseberries despite all that work.

These results show three possible alternatives. Grow just blackcurrant (60 bushes), or grow just gooseberries (100 bushes) or grow a mixture of 30 blackcurrant and 50 gooseberries. Figure 14.5 shows the linear relationship between *B* and *G* and shows that there are many alternatives open to the gardener. This example will be discussed further in Chapter 15.

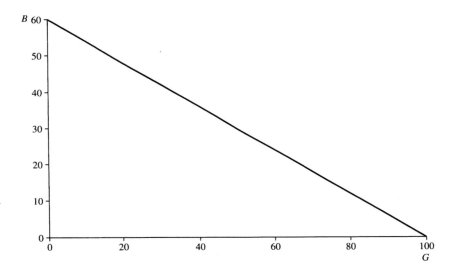

Figure 14.5. Numbers of blackcurrant and gooseberry bushes in 300 m²

KEY POINT	An equation is still true as long as both sides are altered in the same way. The usual steps are:

- to add the same value to both sides,
- to subtract the same value from both sides,
- to multiply both sides by the same value.

Worked example

14.3   If $5x + 7y = 114$,

(a) find x if $y = 2$
(b) find y if $x = 6$
(c) find an equation for y in the form $y = a + bx$
(d) find the gradient, i.e. the rate of change of y with respect to x

Solution   (a) If $y = 2$, then $5x + 7(2) = 114$

$$5x + 14 \quad = 114$$

Subtract 14 from both sides:

$$5x = 100$$

Divide both sides by 5:

$$x = \tfrac{100}{5} = 20$$

$\therefore$ If $y = 2$, $x = 20$

(b) If $x = 6$, then $5(6) + 7y = 114$

$$30 + 7y = 114$$

Subtract 30 from both sides:

$$7y = 84$$

Divide both sides by 7:

$$y = \tfrac{84}{7} = 12$$

$\therefore$ If $x = 6$, $y = 12$

(c) $5x + 7y = 114$

Subtract $5x$ from both sides:

$$7y = 114 - 5x$$

Divide both sides by 7:

$$y = \tfrac{114}{7} - \tfrac{5}{7}x$$

Thus either

$$y = \tfrac{114}{7} - \tfrac{5}{7}x$$

or $y = 16.29 - 0.71x$ (accurate to two decimal places.)

(d) The gradient is b, and so the answer is $-\tfrac{5}{7}$ or -0.71.

Worked example

14.4   Solve the following equation:

$$7(x - 3) = 10 + 4(x - 1)$$

Solution   First multiply out the brackets, giving

$$7x - 21 = 10 + 4x - 4$$

Add 21 to both sides:

$$7x = 10 + 4x - 4 + 21$$

i.e. $7x = 4x + 10 - 4 + 21$

$$7x = 4x + 27$$

Subtract $4x$ from both sides:

$$3x = 27$$

Divide both sides by 3:

$$x = 9$$

Self-assessment questions 14.2

1. For the equation discussed at the start of this chapter, $R = 8.45L$, what value does R take if L is 8?

2. In the linear equation $R = 8.45L$ what is the intercept? (Trick question?)

3. If $y = 8 - 2x$ find x if $y = 0$.

Exercise 14.2

1. Sketch $y = 4 + 3x$ for x between -2 and $+5$. Find and show on your sketch the intercept and the gradient.

2. Rearrange $3(2x + y) = 10$ into the form $y = a + bx$ and hence state the gradient.

3. If $P = 12 + 2q$ find q when $P = 36$.

4. Solve the equation $3y + 8 = 2y + 1$.

5. Solve the equation $5(t - 2) = 6(1 - 2t)$.

Test and assignment exercises 14

1. A firm has determined that the costs of ordering are given by a fixed cost independent of the number of orders placed, and a variable cost proportional to the number of orders placed. When 10 are ordered, the total cost is £105, and when 20 are ordered the total cost is £205. If x represents the number ordered and C the total cost (£), show that the linear equation $C = 5 + 10x$ fits the given data and hence
   (a) identify the fixed costs,
   (b) evaluate the cost of ordering 5,
   (c) find how many could be ordered for a cost of £255.

2. If $y = 3x + 7$
   (a) find y if $x = 2$,
   (b) find x if $y = 7$.

3. If $y = 4x - 5$, which one of the following is true?
   (a) $x = 4y - 5$
   (b) $x = 4y + 5$
   (c) $x = \frac{1}{4}y + 5$
   (d) $x = \frac{1}{4}y + \frac{5}{4}$

4. The two variables p and q are linearly related. When $q = 0$, $p = 4$, and when $q = 3$, $p = 19$. Find an equation of the form $p = a + bq$ that fits this information.

5. The variables C and t are linearly related. When $t = 1$, $C = 9$ and when $t = 3$, $C = 35$. Find an equation of the form $C = a + bt$ that fits this information. (You may need to draw a graph.)

6. If $2x + 3y = 60$,
   (a) find x if $y = 10$,
   (b) find y if $x = 9$,
   (c) find an equation for y in the form $y = a + bx$,
   (d) find the gradient, i.e. the rate of change of y with respect to x.

7. Solve $(x + 2)3 - 4 = 9x - 10$.

8. Solve $2(x - 1) = 3(x - 2)$.

9. Solve $\dfrac{2}{x - 3} = \dfrac{3}{x - 2}$.

10. Solve $\dfrac{x - 3}{5} = \dfrac{17 - x}{2}$.

15 Simultaneous equations

Objectives	After reading this chapter, you should be able:

- to formulate suitable problems as a pair of simultaneous equations
- to solve simple simultaneous equations using both graphical and algebraic approaches

15.1 Introduction

Consider again the gardener who wants to plant an area of $300\,\text{m}^2$ with gooseberry and blackcurrant bushes. This situation was first discussed in Chapter 14, where a linear equation connecting the two variables G and B, the numbers of gooseberry and blackcurrant bushes, was derived. This equation arose because each gooseberry bush needs $3\,\text{m}^2$ of land and each blackcurrant bush needs $5\,\text{m}^2$, and all the available $300\,\text{m}^2$ of land is to be used. This produces the linear equation

$$3G + 5B = 300 \qquad\qquad [1]$$

If this equation is referred to again, it can be called equation [1], and a similar notation will be used for future equations in this chapter. The graph of equation [1] is a straight line passing through the points $(G = 0,\ B = 60)$, $(G = 100,\ B = 0)$ and $(G = 50,\ B = 30)$ (see Figure 14.5). Clearly there are many different pairs of values for G and B that satisfy equation [1].

Now besides being restricted to $300\,\text{m}^2$, the gardener has just £120.00 to spend on fruit bushes, with each gooseberry bush costing £1.00 and each blackcurrant bush costing £3.00. This cost information means that there is a second linear equation connecting G and B which is

$$G + 3B = 120 \qquad\qquad [2]$$

The graph of equation [2] is also a straight line and again there are many different pairs of values for G and B that satisfy it. The important question for the gardener is whether he can find a pair of values for G and B such that both equations are true, i.e. all his land and cash are exactly used up. Stated mathematically, this problem is *to solve the two simultaneous equations:*

$$3G + 5B = 300 \tag{1}$$

$$G + 3B = 120 \tag{2}$$

They are called simultaneous because any solution, i.e. a pair of values for G and B, must satisfy them both simultaneously. Such simultaneous equations can be solved either graphically or algebraically, and both methods of solution are given here.

KEY POINT With two unknowns we need two different linear equations to find the two values of these unknowns.

15.2   Graphical approach

Figure 15.1 shows both equation [1] and equation [2] drawn on the same set of axes. To do this the scale on both axes has to be such that both lines will fit on the graph, and this may require a little trial and error. It can be seen from these graphs that at only one point, A, are both equations satisfied by the same values for G and B. The values for G and B at this intersection point can be read off and they provide approximate answers to the original problem of solving a pair of simultaneous equations. Hence the solution is $G = 75$, $B = 15$, i.e. buy 75 gooseberry bushes and 15 blackcurrant bushes. You can verify that

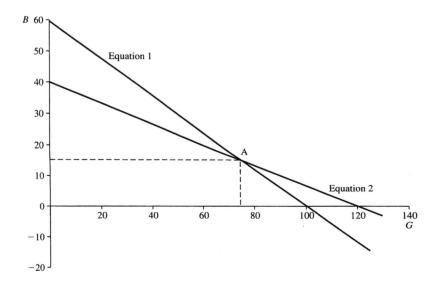

Figure 15.1.
Numbers of gooseberry
and blackcurrant bushes

with these values for G and B both equation [1] and equation [2] are satisfied exactly:

$$3G + 5B = 3(75) + 5(15) = 300$$
$$G + 3B = 75 + 3(15) = 120$$

The graphical approach can only give approximate values for the two variables. Values must be chosen for the variables so that the two lines cross in a region where the results can be easily read from the two axes. This implies a certain amount of planning.

Self-assessment questions 15.2

1. If, having drawn the graph of two equations, they did not cross, what might you try next?

2. Draw the graphs

$$y = 3x + 20$$
$$y = x + 40$$

over the range of x between 0 and 20 and find values for x and y which satisfy both equations.

Exercise 15.2

1. Why could the following pair of simultaneous equations not be solved?

$$2x + 3y = 5$$
$$6x + 9y = 15$$

(Hint: sketch them.)

2. Use a graphical approach to solve

$$4x + 7y = 2135$$
$$6x + 5y = 2075$$

Now compare your answers with those shown later in worked example 15.1 which is the same problem solved algebraically.

3. Draw the graphs

$$y = 5x + 75$$
$$y = 3x + 125$$

over the range of x between 0 and 50 and identify the point of intersection. Show that this point satisfies both equations.

15.3   Algebraic approach

This graphical method is not only approximate, but can be time-consuming, and the following algebraic method is preferred, being accurate and quick once the basic techniques are mastered. Since each equation has two variables in it, it is not possible to solve either individually, but if a new equation could be constructed that had only one variable, then this could be solved using the methods of Chapters 13 and 14.

This is the basic approach, and the first aim is to create a pair of equations from the original two so that in these new equations the coefficients of B are the same (or the coefficients of G are the same). Refer to equations [1] and [2]. To make the G terms the same, leave [1] as it is and multiply both sides of equation [2] by 3. To make the B terms the same it would mean multiplying both sides of [1] by 3 and both sides of [2] by 5, which works, but is not as convenient as the first method. Here is the formal solution for solving these two equations, which are given again for convenience:

$$3G + 5B = 300 \qquad\qquad [1]$$

$$G + 3B = 120 \qquad\qquad [2]$$

Multiply [2] by 3:

$$3G + 9B = 360 \qquad\qquad [3]$$

Rewrite [1]:

$$3G + 5B = 300 \qquad\qquad [1]$$

Subtract [1] from [3]:

$$0 + 4B = 60$$

Divide both sides by 4:

$$B = \tfrac{60}{4} = 15$$

Substitute $B = 15$ into [1]:

$$3G + 5(15) = 300$$

$$3G + 75 = 300$$

Subtract 75 from both sides

$$3G = 225$$

Divide both sides by 3:

$$G = \tfrac{225}{3} = 75$$

The solution is $B = 15$, $G = 75$, which is the same as that obtained graphically. The above method is typical of the solution of all simultaneous linear equations and the following are some comments on it:

1. After inspection of [1] and [2] it was decided to eliminate G rather than B, although elimination of B would produce the same solution.

2. Subtracting [1] from [3] is the key step. Since by definition both sides of any equation are equal, this is just the same as subtracting the same thing from both sides of equation [3].

3. Having found a value for B, G can be found by substituting this value for B into either of the two original equations. Had [2] been used, this would produce

$\therefore$ $G + 3(15) = 120$

$\therefore$ $G + 45 = 120$

$\therefore$ $G = 75$, as before

4. Having obtained a solution, it can be checked by verifying that both of the original equations are satisfied. Substituting $G = 75$ and $B = 15$ into [1] gives:

$3(75) + 5(15) = 225 + 75 = 300$

Substituting $G = 75$ and $B = 15$ in [2] gives:

$75 + 3(15) = 75 + 45 = 120$

KEY POINT

Always write down what you have done at each step. This will allow you to check your work.

Give each newly formed equation a number so that it can be referred to.

Worked example

15.1 At a party held in a public house, the first round bought cost £21.35 and was for four pints of bitter and seven pints of lager. The second round was for six pints of bitter and five pints of lager and cost £20.75. (Two people found the lager too gassy!) What are the prices per pint of bitter and lager?

Solution Let x and y be the price, in pence, of a pint of bitter and lager respectively. Each round then allows an equation to be made, using pence as the standard unit, not pounds.

Round 1 $4x + 7y = 2135$ [1]

Round 2 $6x + 5y = 2075$ [2]

It would be a lot easier just to ask for the price.

We need to solve these simultaneous equations to find x and y. (An algebraic solution is presented here, but you may like to try it using a graph.) The easiest variable to eliminate is x, and although this could be done by multiplying [1] by 6 and [2] by 4, a neater way is to multiply [1] by 3 and [2] by 2.

Multiply [1] by 3:

$$12x + 21y = 6405 \qquad\qquad [3]$$

Multiply [2] by 2:

$$12x + 10y = 4150 \qquad\qquad [4]$$

Subtract [4] from [3]

$$11y = 2255$$

Divide both sides by 11:

$$y = \tfrac{2255}{11} = 205$$

Substitute $y = 205$ into [1]:

$$4x + 7(205) = 2135$$

$$4x + 1435 = 2135$$

Subtract 1435 from both sides

$$4x = 700$$

Divide both sides by 4:

$$x = \tfrac{700}{4} = 175$$

Thus, bitter is £1.75/pint, lager is £2.05/pint. This is not a cheaper pub than the Three Crowns discussed in Chapter 1! Check by substitution into equations [1] and [2].

Into [1]:

$$4(175) + 7(205) = 700 + 1435 = 2135$$

Into [2]:

$$6(175) + 5(205) = 1050 + 1025 = 2075$$

Worked example

15.2 Solve the following simultaneous equations by first eliminating y:

$$2x - 4y = -12 \qquad\qquad [1]$$

$$3x + 2y = 22 \qquad\qquad [2]$$

Solution   Multiply [2] by 2:

$$6x + 4y = 44 \qquad\qquad [3]$$

Rewrite [1]:

$$2x - 4y = -12 \qquad\qquad [1]$$

Add [3] to [1]:

$$8x = 32$$

Divide both sides by 8:

$$x = 4$$

Substitute $x = 4$ into [1]:

$$2(4) - 4y = -12$$
$$8 - 4y = -12$$

Subtract 8 from both sides:

$$-4y = -12 - 8$$
$$-4y = -20$$

Divide both sides by -4

$$y = \frac{-20}{-4} = +5$$

Solution: $x = 4$, $y = 5$.

If any of the manipulations with negative signs in this solution are not clear, read Chapter 1.

Worked example

15.3   Solve the following simultaneous equations (same as in worked example 15.2) by first eliminating x:

$$2x - 4y = -12 \qquad\qquad [1]$$
$$3x + 2y = 22 \qquad\qquad [2]$$

Solution   Multiply [1] by 3:

$$6x - 12y = -36 \qquad\qquad [3]$$

Multiply [2] by 2:

$$6x + 4y = 44 \qquad\qquad [4]$$

Subtract [4] from [3]

$$-12y - 4y = -36 - 44$$
$$\therefore \quad -16y = -80$$

Divide both sides by -16:

$$y = \frac{-80}{-16} = +5$$

Substitute $y = +5$ into [1]:

$$2x - 4(5) = -12$$

$$2x - 20 = -12$$

Add 20 to both sides:

$$2x = -12 + 20$$

$$2x = 8$$

Divide both sides by 2:

$$x = 4$$

Solution: As in worked example 15.2, $x = 4$ and $y = 5$.

Worked example

15.4   Solve for x and y:

$$3x + 8 = 2y \qquad [1]$$

$$5y = 1 - 2x \qquad [2]$$

Solution   These two equations are not in the same form as those previously solved, and so the first step is to rewrite them.

$$3x + 8 = 2y \qquad [1]$$

Subtract 8 from both sides:

$$3x = 2y - 8$$

and subtract $2y$ from both sides

$$3x - 2y = -8 \qquad [3]$$

Equation [3] is now in the form we are used to. Now to put [2] into the same form:

$$5y = 1 - 2x \qquad [2]$$

Add $2x$ to both sides:

$$2x + 5y = 1 \qquad [4]$$

Equations [3] and [4] can now be solved as before.

Multiply [4] by 3:

$$6x + 15y = 3 \qquad [5]$$

Multiply [3] by 2:

$$6x - 4y = -16 \qquad [6]$$

Subtract [6] from [5]:

$$15y - (-4y) = 3 - (-16)$$
$$15y + 4y = 3 + 16$$
$$19y = 19$$
$$\therefore \quad y = 1$$

Substitute $y = 1$ into [1]

$$3x + 8 = 2(1)$$
$$3x + 8 = 2$$

Subtract 8 from both sides:

$$3x = 2 - 8$$
$$3x = -6$$

Divide both sides by 3:

$$x = -6/3 = -2$$

Solution $x = -2$, $y = 1$.

Self-assessment questions 15.3

1. Eliminate x and hence solve

$$x + 3y = 7$$
$$x - y = -1$$

2. Equations [1] and [2] have been used to produce equation [3]. Write down the missing explanation.

$$2x + 4y = 6 \qquad [1]$$
$$5x - 2y = 13 \qquad [2]$$
$$12x = 32 \qquad [3]$$

Exercise 15.3

1. Solve:

$$4x - y = 9$$
$$5x + 2y = 21$$

2. Solve:

$$6p + 2q = 38$$
$$6q - 5p = 22 \text{ (Take care!)}$$

3. Solve:

$$4r + 2s = 56$$
$$r + 5s = 50$$

4. Solve:

$$3a + 2b = 80$$
$$5a + b = 75$$

Test and assignment exercises 15

1. When some items are offered for sale, the quantity demanded (q) by customers will decrease if the price (p) increases. However, as the price increases the quantity supplied by the retailer will increase. For a particular item the demand and supply equations are:

    Demand: $q + 2p = 1000$

    Supply: $q - 2p = -200$

    (a) Graph both these on the same axes and find the values of q and p (in pence) which make supply equal demand.
    (b) Solve these equations algebraically to confirm your solution from (a). (Eliminate p first.)

2. Solve the following simultaneous equations by first eliminating B, and confirm that you get the same solution as that given in the text:

    $3G + 5B = 300$

    $G + 3B = 120$

3. The sum of two numbers is 20 and their difference is 4. Write this as two simultaneous equations and hence find the two numbers.

4. Solve for x and y:

    $3x - 4y = 5$

    $2x + 3y = 26$

5. Solve for p and q:

    $2p - 3q = 5$

    $5p + 2q = 22$

6. Solve for a and b:

    $-3a + 4b = 5$

    $2a - 5b = -15$

7. Solve for x and y:

    $7x - 2y = 1$

    $-2x + 7y = 82$

8. Solve for x and y:

    $4x = 3y + 10$

    $-x = -2y + 10$

9. Solve for x and y:

    $x + 6 = y$

    $3y - 16 = 4x$

10. Solve for x and y:

    $4x = 7 + 3y$

    $-6y + 8x = 15$

16 Quadratic equations

Objectives

After reading this chapter, you should be able:

- to understand the behaviour of quadratic functions
- to solve quadratic equations by both factorization and the standard formula method

16.1 Introduction

Most of the equations in previous chapters have been linear; that is, their graphs have been straight lines. Many mathematical relationships are not linear and are therefore called non-linear. The quadratic is the commonest and simplest of these non-linear relationships. The example from Chapter 13

$$A = \pi r^2$$

is an example of a quadratic relationship. The usual form for a linear equation is

$$y = a + bx$$

whereas the usual form for a quadratic is

$$y = ax^2 + bx + c$$

where a and b are coefficients and c is a constant. Occasionally b or c are zero but if a is zero the equation would become a linear equation.

KEY POINT

A quadratic equation takes the form

$$y = ax^2 + bx + c$$

where a, b and c are numbers; for example

$$y = 2x^2 - 5x + 17$$

16.2 Solving quadratic equations

Notice that the main difference between linear and quadratic equations is the presence of the x^2 term or x to the power 2.

For example, when a cricket ball is hit over a flat pitch, the

Never mind the
quadratic, CATCH IT!

relationship between the height of the ball above the ground, H metres, and the time the ball has been in the air, t seconds, can be shown to be (approximately):

$$H = 5t(5 - t)$$

It can be seen from this equation that when $t = 0$, $H = 0$, i.e. initially the ball is at ground level, while if $t = 5$, $H = 0$ again, showing that after a flight lasting 5 s, the ball is again at ground level. The results of evaluating H for values of t ranging from 0 to 5 are shown in Table 16.1, and the graph of these results is shown in Figure 16.1. This curve is a typical example of a *parabola* or parabolic curve.

Table 16.1.
Height and time for a
cricket ball

Time t (s)	0	1	2	3	4	5
Height H (m)	0	20	30	30	20	0

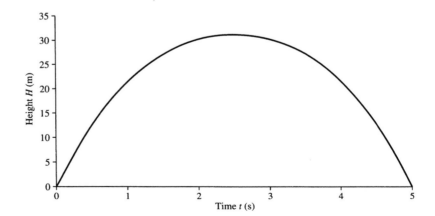

Figure 16.1.
Height and time for a
cricket ball

Consider the problem of trying to find when the ball is at a height of 20 m. From either Table 16.1 or Figure 16.1 it can be seen that the ball is at a height of 20 m twice, first when $t = 1$ and then $t = 4$. When is the ball at a height of 25 m? This is not clear from the table, but the graph does allow two approximate answers to be obtained, the first after just less than 1.5 s and then after slightly more than 3.5 s. In this chapter, more accurate methods of obtaining these results are discussed and to illustrate these methods the problem of finding t when $H = 20$ will be solved, although the answer of $t = 1$ or $t = 4$ is already known. (The problem when $H = 25$ will be solved later as an example.)

The original equation with $H = 20$ is

$$20 = 5t(5 - t)$$

and after multiplying out the brackets, this becomes

$$20 = 25t - 5t^2$$

Add $5t^2$ to both sides:

$$5t^2 + 20 = 25t$$

Subtract $25t$ from both sides:

$$5t^2 - 25t + 20 = 0$$

Divide both sides by 5:

$$t^2 - 5t + 4 = 0 \qquad [1]$$

This equation is a typical example of a quadratic equation, i.e. an equation where the most complicated term is the variable squared.

The graphical method requires the graph of

$$y = t^2 - 5t + 4$$

to be drawn, and then this graph is used to read off the values of t for which $y = 0$. By evaluating y for various values of t, the graph in Figure 16.2 is produced, and this clearly shows that when $y = 0$, $t = 1$ or $t = 4$, which confirms the earlier solutions. Any quadratic equation like equation [1] will allow a graph to be drawn, and the solutions will be where the curve crosses the horizontal axis. There are three possibilities:

1. Two distinct solutions, as in Figure 16.2 and Figure 16.3(a).
2. One single solution, as in Figure 16.3(b).
3. No solution, as in Figure 16.3(c).

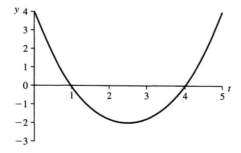

Figure 16.2.
Equation [1]

Solving equation [1] should produce the two solutions already known, $t = 1$ and $t = 4$, and three methods of solving this and similar equations are now discussed.

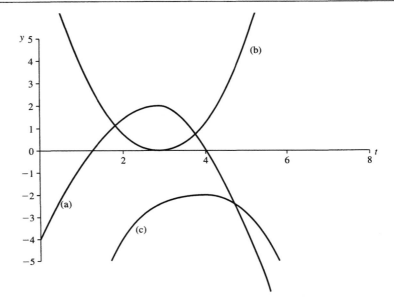

Figure 16.3.
Three possible types of
solution

KEY POINT

Quadratic equations when sketched have a parabolic shape.

They either have two different solutions, no solutions or in some cases just one solution

This graphical method, although instructive, can only produce approximate solutions, and one of the following two algebraic methods should be used.

Equation [1] can be solved by rewriting it in terms of its two factors. Since

$$t^2 - 5t + 4 = (t - 1)(t - 4)$$

equation [1] is

$$(t - 1)(t - 4) = 0$$

If you find this step difficult it may be sensible to check that it is correct by multiplying $(t - 1)$ by $(t - 4)$ using the rules of section 13.3.

Whenever the product of two terms is zero, then either the first or the second (or both) must be zero. Hence either

$$(t - 1) = 0 \quad \text{or} \quad (t - 4) = 0$$

either

$$t = 1 \quad \text{or} \quad t = 4$$

Not all quadratic equations can be factorized, and even when they can it is not always easy to obtain the two factors. It is really a matter of trial and error, testing each possibility suggested. For example, consider factorizing and hence solving

$$x^2 + 3x - 10 = 0 \qquad [2]$$

The only way x^2 can be split into its factors is by multiplying x and x, while -10 could be either $(-10,1)$, $(10,-1)$,$(-5,2)$ or $(5,-2)$. This means that there are four possibilities for the two factors of [2]:

$$(x - 10)(x + 1) = x^2 + x - 10x - 10 = x^2 - 9x - 10 \text{ (no)}$$
$$(x + 10)(x - 1) = x^2 - x + 10x - 10 = x^2 + 9x - 10 \text{ (no)}$$
$$(x - 5)(x + 2) = x^2 + 2x - 5x - 10 = x^2 - 3x - 10 \text{ (no)}$$
$$(x + 5)(x - 2) = x^2 - 2x + 5x - 10 = x^2 + 3x - 10 \text{ (yes)}$$

Thus the factors of [2] are

$$(x + 5)(x - 2) = 0$$

either

$$x + 5 = 0 \quad \text{or} \quad x - 2 = 0$$

either

$$x = -5 \quad \text{or} \quad x = 2$$

The third and final method for solving quadratics is to apply the following formula, which can be proved to give the solutions, if there are any, of any quadratic equation. When the quadratic equation is written in the form

$$ax^2 + bx + c = 0$$

where a, b and c are constants with $a \neq 0$, then the solution is

$$x = \frac{-b \pm \sqrt{(b^2 - 4ac)}}{2a}$$

(The plus or minus sign, $\pm$, will allow two solutions to be found.) When equation [1] is written in this form, $a = 1$, $b = -5$ and $c = 4$, and so the solution is (note that in [1] t is the variable, not x, but this has no important effect):

$$t = \frac{-(-5) \pm \sqrt{[(-5)^2 - 4(1)(4)]}}{2(1)}$$

i.e.

$$t = \frac{+5 \pm \sqrt{25 - 16}}{2}$$

i.e.

$$t = \frac{+5 \pm \sqrt{9}}{2}$$

i.e.

$$t = \frac{+5 \pm 3}{2}$$

Thus either

$$t = \frac{+5 + 3}{2} \quad \text{or} \quad t = \frac{+5 - 3}{2}$$

either

$$t = \frac{8}{2} \quad \text{or} \quad t = \frac{2}{2}$$

either

$$t = 4 \quad \text{or} \quad t = 1$$

KEY POINT

The solution of $ax^2 + bx + c = 0$ is

$$x = \frac{-b \pm \sqrt{(b^2 - 4ac)}}{2a}$$

Worked example

16.1 A small rectangular lawn is to be laid with a total area of $40\,\text{m}^2$ and with the length $3\,\text{m}$ more than the width. Find the width.

Solution Let the width be x, and so the length must be $x + 3$ and the area must be $x(x + 3)$. Thus

$$x(x + 3) = 40$$

i.e.

$$x^2 + 3x = 40$$

i.e.

$$x^2 + 3x - 40 = 0 \qquad [3]$$

Equation [3] is a quadratic. Trying to find the factors produces

$$x^2 + 3x - 40 = (x + 8)(x - 5)$$

Hence

$$(x + 8)(x - 5) = 0$$

∴ Either

$$x + 8 = 0 \quad \text{or} \quad x - 5 = 0$$

so

$$x = -8 \quad \text{or} \quad x = 5$$

There is only one solution, $x = 5$, since $x = -8$ is clearly not practicable. If finding the factors is too difficult, then the formula can always be used, and in this case $a = 1$, $b = 3$, $c = -40$, giving

$$x = \frac{-3 \pm \sqrt{[(3)^2 - 4(1)(-40)]}}{2(1)}$$

i.e. $$x = \frac{-3 \pm \sqrt{[9 + 160]}}{2}$$

i.e. $$x = \frac{-3 \pm \sqrt{169}}{2}$$

$$x = \frac{-3 \pm 13}{2}$$

∴ Either

$$x = \frac{-3 + 13}{2} \quad \text{or} \quad x = \frac{-3 - 13}{2}$$

thus

$$x = \frac{10}{2} \quad \text{or} \quad x = \frac{-16}{2}$$

and so

$$x = 5 \quad \text{or} \quad x = -8$$

Again the solution is $x = 5$.

Worked example

16.2   When a car traveling at v km/hour brakes to a stop on a dry road, the approximate distance taken to stop, D metres, which includes the thinking time of the driver, is given by:

$$D = \frac{v}{5} + \frac{v^2}{170}$$

This equation gives a better approximation than the one discussed in Chapter 13. If in an accident a car travelled 60 metres before stopping, estimate the speed that the car must have been doing before the accident.

Solution   We need to solve

$$60 = \frac{v}{5} + \frac{v^2}{170}$$

Multiply both sides by 170:

$$10\,200 = 34v + v^2$$

Subtract 10 200 from both sides:

$$0 = 34v + v^2 - 10\,200$$

Rearrange:

$$v^2 + 34v - 10\,200 = 0$$

Solving this equation by finding the factors is clearly at best difficult, and not worth the time. In this case, using the formula is a more direct method, with $a = 1$, $b = 34$ and $c = -10\,200$.

$$v = \frac{-34 \pm \sqrt{[(34)^2 - 4(1)(-10\,200)]}}{2(1)}$$

$$v = \frac{-34 \pm \sqrt{1156 + 40\,800}}{2}$$

$$v = \frac{-34 \pm \sqrt{41\,956}}{2}$$

$$v = \frac{-34 \pm 204.83}{2}$$

So either

$$v = \frac{-34 - 204.83}{2} \quad \text{or} \quad v = \frac{-34 + 204.83}{2}$$

either

$$v = \frac{-238.83}{2} \quad \text{or} \quad v = \frac{170.83}{2}$$

either

$$v = -119.41 \quad \text{or} \quad v = 85.41$$

Hence $v = 85.4$ km/hour. The negative answer is clearly not practicable, and since one decimal place is enough accuracy in a final answer for speed, two decimal places were used in the solution.

Worked example

16.3   Solve

$$x^2 - 6x + 9 = 0$$

Solution   This equation can be factorized, giving

$$x^2 - 6x + 9 = (x - 3)(x - 3)$$

and so

$$(x - 3)(x - 3) = 0$$

either

$$x - 3 = 0 \quad \text{or} \quad x - 3 = 0$$

i.e. either

$$x = 3 \quad \text{or} \quad x = 3$$

This is an example of a quadratic equation with a single solution (see Figure 16.3(b)), although strictly it should be called a repeated solution.

Worked example

16.4   After how many seconds will the cricket ball discussed earlier be at a height of 25 m?

Solution   Using the equation at the start of this chapter with $H = 25$ gives:

$$25 = 5t(5 - t)$$

i.e. $25 = 25t - 5t^2$

Divide both sides by 5:

$$5 = 5t - t^2$$

Add t^2 to both sides:

$$t^2 + 5 = 5t$$

Subtract $5t$ from both sides:

$$t^2 - 5t + 5 = 0$$

This quadratic cannot be factorized simply, so the formula should be used with $a = 1$, $b = -5$, and $c = 5$, giving

$$t = \frac{-(-5) \pm \sqrt{[(-5)^2 - 4(1)(5)]}}{2(1)}$$

i.e. $$t = \frac{+5 \pm \sqrt{(25 - 20)}}{2}$$

i.e. $$t = \frac{+5 \pm \sqrt{5}}{2}$$

$$t = \frac{5 \pm 2.236}{2}$$

either

$$t = \frac{7.236}{2} \quad \text{or} \quad t = \frac{2.764}{2}$$

either $t = 3.618$ or $t = 1.382$

The ball reaches a height of 25 m after 1.38 seconds and again after 3.62 seconds.

(Compare these results with the approximate ones obtained by a graphical method earlier.)

Worked example

16.5   Solve

$$x^2 + x + 1 = 0$$

Solution   This will not factorize, so use the equation with $a = 1$, $b = 1$ and $c = 1$, giving

$$x = \frac{-1 \pm \sqrt{[(1)^2 - 4(1)(1)]}}{2(1)}$$

i.e.

$$x = \frac{-1 \pm \sqrt{1 - 4}}{2}$$

i.e.

$$x = \frac{-1 \pm \sqrt{-3}}{2}$$

Now the square root of a negative number does not exist, so there are no values of x that satisfy this equation. This situation can often occur, and can be understood best by considering a graphical attempt to find a solution in which the curve does not cross or touch the horizontal axis, as in Figure 16.3(c).

KEY POINT

The three possibilities for solving a quadratic

$$ax^2 + bx + c = 0$$

can now be described by:

1. Two distinct solutions, as in Figure 16.3(a), when $b^2 - 4ac$ is positive.
2. One repeated solution, as in Figure 16.3(b), when $b^2 - 4ac$ is zero.
3. No solution, as in Figure 16.3(c), when $b^2 - 4ac$ is negative.

Worked example

16.6   Solve

$$x^2 - 3x = 0$$

Solution   By factoring,

$$x^2 - 3x = x(x - 3)$$

and so

$$x(x - 3) = 0$$

$\therefore$ either $x = 0$ or $x - 3 = 0$

$\therefore$ either $x = 0$ or $x = 3$

Do not ignore the solution $x = 0$ in problems like this.

Self-assessment questions 16.2

1. Which of the following are quadratic equations?
   (a) $y = 5 - 2x^2$ (b) $y = 2x(3 - x)$
   (c) $y = 3x(7 - 2)$ (d) $t^2 = 6n + 8$

2. Does $(2x - 2)(x + 6) = 2x^2 + 10x - 12$?

3. Is this the correct formula for the solution of $ax^2 + bx + c = 0$?

$$x = \frac{-b \pm \sqrt{(c^2 - 4ab)}}{2a}$$

Exercise 16.2

1. Solve by factorizing:
   $$3x^2 - 10x + 8 = 0$$

2. Solve by factorizing:
   $$x^2 - 4x + 4 = 0$$

3. Solve:
   $$2x^2 + 2x - 7 = 0$$

4. Solve the following quadratic equation:
   $$6x^2 - 11x - 10 = 0$$

5. Explain why the following equation cannot be solved:
   $$2x^2 + 2x + 3 = 0$$

Test and assignment exercises 16

1. Draw the graph of $y = 2x^2 - 11x + 12$ for x between 0 and 6 and hence estimate the values of x that satisfy the quadratic equation

$$2x^2 - 11x + 12 = 0$$

2. By factorizing $x^2 - 3x + 2$ solve the equation

$$x^2 - 3x + 2 = 0$$

3. By using an appropriate formula solve

$$x^2 - 3x - 3 = 0$$

4. Obtain an accurate solution to question 1.

5. Use an alternative method to solve question 2, so confirming the previous solution.

6. Use factorization to solve the following equations:
(a) $x^2 + 3x + 2 = 0$
(b) $p^2 - 3p = 0$
(c) $6x^2 - 13x + 6 = 0$
(d) $3n^2 + 11n - 4 = 0$
(e) $-2t^2 - t + 21 = 0$

7. Use an appropriate formula to solve, where possible, the following equations:
(a) $x^2 + x - 1 = 0$
(b) $t^2 + 3t + 5 = 0$
(c) $5p^2 + 5p - 5 = 0$
(d) $0.73x^2 + 1.03x - 0.52 = 0$
(e) $-2n^2 + 3n + 3 = 0$

8. Simplify and then solve, where possible, the following equations:
(a) $x(x - 2) = 1$
(b) $2t(t + 3) = t^2 + 4$
(c) $\sqrt{x} = x - 10$
(d) $(n + 2)n = (n - 2)(3 - n)$
(e) $\dfrac{3}{n} = n + 2$

Solutions

Self-assessment question 1.2

1. (a) $24 - 8 = 16$
   (b) $6 + 2 - 5 = 3$
   (c) $30 - 9 = 21$
   (d) $12 - 8 = 4$
   (e) $(18 - 4) \times 3 = 14 \times 3 = 42$
   (f) $2 \times 3 = 6$

Exercise 1.2

1. $5 \times 0.42 = £2.10$

2. (a) 1304
   (b) 263
   (c) 9
   (d) 186

3. $2(16) + 3(14) + 3(11) = 107$

4. (a) $4 \times (12 - 5) + 2$
   (b) $3 \times (11 + 6 - 2)$
   (c) $(9 \times 10) - 4 + 6$

5. $21\,840 \div (13 \times 35) = 48$

Self-assessment question 1.3

1. Positive

Exercise 1.3

1. (a) $45 + 1 = 46$
   (b) $104 \div 2 = 52$
   (c) $40 \div 4 = 10$

2. $36 - 40 = -£4$ (£4 overdrawn)

3. (a) 4
   (b) -16
   (c) -4
   (d) -16
   (e) 16

4. $75 - (-53) = £128$

5. (a) -45
   (b) 45
   (c) 5
   (d) -5

6. $60 - (-120) = 180\,\text{m}$

Self-assessment question 1.4

1. Simplicity, accuracy

Exercise 1.4

1. 6000, 9000

2. (a) 5560
   (b) 147 000
   (c) 1130

3. (a) 200
   (b) 534 900
   (c) 600
   (d) 0

4. (a) £3 873 600
   (b) £3 870 000

5. 470

Self-assessment question 2.2

1. $\frac{1}{2}, \frac{7}{12}, \frac{3}{5}, \frac{5}{8}$

Exercise 2.2

1. $\frac{3}{5} \times 12\,500 = £7500$

2. (a) $\frac{9}{12} + \frac{8}{12} = 1\frac{5}{12}$
   (b) $\frac{21}{24} - \frac{4}{24} = \frac{17}{24}$
   (c) $1\frac{1}{3} + 1\frac{5}{6} = 3\frac{1}{6}$

3. Copper $\frac{7}{10} \times 750 = 525\,\text{g}$
   Tin $\frac{9}{50} \times 750 = 135\,\text{g}$
   Zinc $\frac{3}{25} \times 750 = 90\,\text{g}$

4. (a) $\frac{5}{6} \times \frac{2}{5} = \frac{1}{3}$
   (b) $\frac{7}{8} \times 2 = 1\frac{3}{4}$
   (c) $\frac{5}{3} \times \frac{6}{5} = 2$

5. Upper 200 pupils
   Middle 450 pupils
   Lower $1200 - 650 = 550$ pupils

Self-assessment questions 2.3

1. $\frac{2}{5}$, $2\frac{12}{100}$

2. (a) $3\frac{1}{2}$
   (b) 3.25

Exercise 2.3

1. (a) 6.407
   (b) 0.474

2. (a) 61.918
   (b) 1.547 17

3. (a) 8.4
   (b) 7.3

4. $872 \div 1.26 = 692$

5. $\frac{7}{8} = 0.875$

Self-assessment question 3.2

1. (a) 3% faulty
   (b) 38% female
   (c) 18% left-handed

Exercise 3.2

1. 25% of £28 = £7

2. $1000 + 100 + 110 = £1210$

3. $\frac{14}{40} \times 100 = 35\%$

4. (a) $\frac{30}{100} \times 80 = 24$
   (b) $\frac{45}{100} \times 120 = 54$
   (c) $\frac{60}{100} \times 45 = 27$

5. $600 \times 1.09 = £654$

6. $18 \div 120 \times 100 = 15$

7. (a) $\frac{3}{5} \times 100 = 60\%$
   (b) $\frac{7}{8} \times 100 = 87.5\%$
   (c) $\frac{13}{20} \times 100 = 65\%$

8. $\frac{4}{16} \times 100 = 25\%$

9. A $\frac{24\,000}{120\,000} \times 100 = 20\%$
   B $\frac{30\,000}{120\,000} \times 100 = 25\%$
   C $100\% - 20\% - 25\% = 55\%$

10. $7.50 \times \left(1 + \frac{17.5}{100}\right) = 8.81$
    $8.81 \times \left(1 + \frac{10}{100}\right) = £9.69$

Self-assessment question 3.3

1. (a) 4 : 1
   (b) 2.5 : 1
   (c) 3 : 1

Exercise 3.3

1. (a) 1 : 2
   (b) 1 : 6
   (c) 3 : 2

2. (a) 8 : 1
   (b) 400 000 : 1
   (c) 1 : 50

3. 40 : 25, 8 : 5

4. $\frac{2}{5} \times 2 = £0.80$

5. $\frac{2}{3} \times 60 = 40$ students

6. $\frac{5}{12} \times 6000 = £2500$

Self-assessment question 4.2

1. $3^{-2} = 1/(3^2) = \frac{1}{9}$, negative power gives reciprocal

Exercise 4.2

1. (a) 64
   (b) 49
   (c) $\frac{1}{9}$
   (d) 10 000

2. (a) $4^5 = 1024$
   (b) $3^{-1} = \frac{1}{3}$
   (c) $25 \times 27 = 675$

3. (a) $100 \times 27 = 2700$
   (b) $9 \times 1 = 9$
   (c) $8 \times 3 = 24$

4. $125 - 121 = 4$

Self-assessment question 4.3

1. $\sqrt{9} = 3$, roots, e.g. power $\frac{1}{2}$ = square root

Exercise 4.3

1. (a) 12
   (b) 22
   (c) 5

2. (a) 192.35
   (b) 0.112

3. (a) $\frac{2}{3}$ (b) $\frac{5}{7}$

4. (a) 4   (b) 6

Self-assessment question 4.4

1. (a) $8000(1.05)^3 = £9\,261$
   (b) $10\,000 \times 9/100[1.09^{15}/(1.09^{15} - 1)]$
      $= £1241$ per year, £103.38 per month

Exercise 4.4

1. $(1.063)^4 \times 12\,500 = £15\,960$
   Interest $= £3460$

2. $(1.065)^5 \times 2500 = £3425$

3. $50\,000 \times 8.5/100[1.085^{25}/(1.085^{25} - 1)]$
   $= £4885.58$ per year, £407.13 per month.
   Total paid over 25 years is
   £407.13 × 25 × 12 = £122 139

4. $20\,000 \times 8.75/100[1.0875^{10}/(1.0875^{10} - 1)]$
   $= £3082$ per year, £256.85 per month

Self-assessment question 4.5

1. (a) 9.6742×10^4
   (b) 8.6×10^{-5}
   (c) 9×10^{-3}

Exercise 4.5

1. (a) 375 000
   (b) 0.000 987
   (c) 0.007 52

2. 8.675×10^{-1}

3. $0.000\,625 + 0.052 = 0.052\,625$

Self-assessment questions 5.2

1.
Score		Frequency
0	⦀⦀⦀ ⦀⦀⦀⦀	9
1	⦀⦀⦀ ⦀	6
2	⦀⦀⦀	3
3	⦀	1
4	⦀	1

2. (a) A cost of £3.00, £5.90 or £15.00 could go
   into two classes.

(b) There is a gap from £9.00 to £10.00. A
   better set of classes would be:

   0.00 to under   3.00
   3.00 to under   5.00
   5.00 to under 10.00
   10.00 to under 15.00
   15.00 and over

The unequal intervals are not an error
but could cause an error if a histogram
were to be drawn.

Exercise 5.2

1.
Class	Frequency
10–29	6
30–49	17
50–69	23
70–89	13
90–109	1
Total	**60**

Either is appropriate but the original is
slightly better because it separates failures out
better if a failure is any mark less than 40%.

2.
Class		Frequency
0 to under 40	⦀⦀⦀⦀ ⦀⦀	7
40 to under 80	⦀⦀⦀⦀ ⦀⦀⦀⦀ ⦀⦀⦀⦀	14
80 to under 120	⦀⦀⦀⦀ ⦀	6
120 to under 160	⦀⦀	2
160 to under 200	–	0
200 or more	⦀	1

3.
Class	Frequency
300–309	2
310–319	5
320–329	9
330–339	11
340–349	9
350–359	2
360–369	2

Self-assessment questions 5.3

1. $3(30.00) + 8.22 + 3 \times 2 \times 0.75$
   $= 90 + 8.22 + 4.50 = £102.72$
2. Remarriages as a % of all
   $= 100(350 - 222)/350$
   $= 100(87)/350$
   $= 36.571\%$

Exercise 5.3

1. Only Girassol Hotel offer all facilities needed.
   Cost $= 3 \times 605 + 3 \times 14(10.90)$
   $$+ 1 \times 14(5.50 + 15.00)$$
   $$= 1815 + 457.80 + 287.00$$
   $$= \pounds 2559.80 \text{ (quite expensive!)}$$

2. Table 5.9 gives 18.851 48 and a calculator gives 18.852 036

3. (a) £121.57
   (b) $133.95 - 121.57 = \pounds 12.38$
   (c) Total $= 15 \times 12 \times 121.57 = \pounds 21\,882.60$
       (Wow!!)
   (d) Total $= 15 \times 12 \times (133.95 - 121.57)$
       $$= \pounds 2\,228.40$$

Self-assessment question 6.2

1. If 'now' is November 1998, start in January 1995, mark the axis in months up to January 1999, to allow for the 'future'

Exercise 6.2

1. Lowest in 1907 (approximately)

2. 1978 (again this is approximate)

3. Clearly seasonal data, high in the colder months.

Self-assessment question 6.3

1. (a) Histogram
   (b) Pie chart
   (c) Bar chart

Exercise 6.3

1. 10%

2. 9% of $4000 = 360$

3. 26% of females and 18% of males means $(26 + 18)/2 = 22\%$ of people.

4. See Table S1 and Figure S1.

   Table S1

Year	First	Second
1961	36	21
1971	54	36
1981	74	61
1991	71	57
1992	75	59

Figure S1. Multiple bar chart for marriages

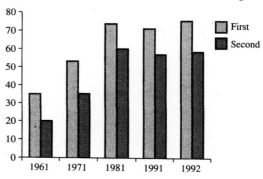

Self-assessment questions 7.2

1. Are you 'normal'?

2. See Figure S2.

   Figure S2. Scatter diagram showing car ownership information

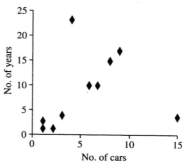

   Except for persons D and G there is a clear relationship.

3. See Figure S3.

   Figure S3. Scatter diagram showing goals scored

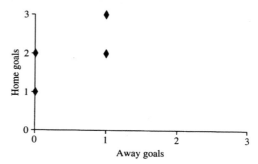

Exercise 7.2

1. See Figure S4.

   Figure S4. Relationship between hip size and shoe size

   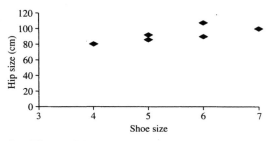

2. See Figure S5.

   Figure S5. Computer shape

3. See Figure S6.

   Figure S6. Scatter diagram showing production costs for books

   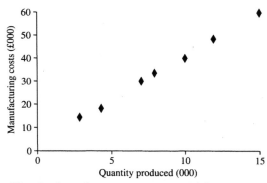

   Clearly there is a very strong positive relationship, as would be expected.

Self-assessment questions 8.2

1. 73 kg

2. 24 months

Exercise 8.2

1. See Figure S7.

Figure S7. Heights and weights of women

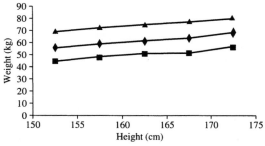

2. (a) See Figure S8.

   Figure S8. Survival patterns for the two sexes

   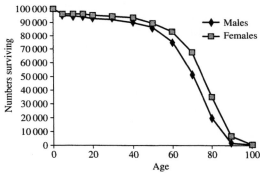

   (b) Women live longer than men
   Most die before 100 years
   Women are less likely to die at an early age

   (c) 90 000 males and 92 000 females

Self-assessment question 8.3

1. Speed is approximately 40 m/min

Exercise 8.3

1. The barge starts to slow after 3 min

2. The maximum gradient (speed) is 132.5 m/min

3. See Figure S9.

   Figure S9. Temperature of a metallic body cooling over time

   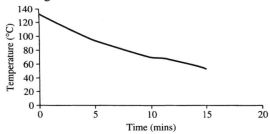

(a) The rate of fall is getting less.
(b) Approximately 50 °C
(c) −7. That is the temperature is falling at a rate of 7 °C per minute

Self-assessment question 9.2

1. Grouped data firstly approximates the data

Exercise 9.2

1. 990/5 = £198
   Total 6 × 199 = £1194
   £1194 − £990 = £204

2. Total 44 years 8 months
   Average 11 years 2 months

3. Total time = 3.5 hours
   Total distance = 50 + 45 = 95 miles
   Average speed = 95 ÷ 3.5 = 27.1 mph

4. Mean = $(172 \times 15 + 175 \times 20 + 180 \times 40$
   $+ 185 \times 35 + 190 \times 17 + 192 \times 13$
   $+ 195 \times 10)/150 = 27431/150$
   $= £182.87$

5. Mean = $(500 \times 24 + 700 \times 50 + 900 \times 64$
   $+ 1100 \times 42 + 1300 \times 20)/200$
   $= 176800/200 = £884$

Self-assessment questions 9.3

1. Advantage – not dependent on extreme values
   Disadvantage – not useful for statistical theory

2. You would expect approximately half the population to be below the average. If the average used was the median then exactly half would lie below (and half above)

Exercise 9.3

1. median = £18 000
   new median = £17 000

2. 6

3. 9 °C

Self-assessment question 9.4

1. Clothes manufacturer making one size only

Exercise 9.4

1. 4

2. Red

3. Size 9

Self-assessment questions 10.2

1. It uses all data values

2. $s = \sqrt{(528.72/6 - 9^2)} = 2.67$

Exercise 10.2

1. Mean = £98
   $s = \sqrt{(48\,850/5 - 98^2)} = \sqrt{166} = £12.88$

2. Season 1: Mean = 21
   $s = \sqrt{(5826/8 - 21^2)} = 16.9$
   Season 2: Mean = 33
   $s = \sqrt{(18\,268/10 - 33^2)} = 27.2$

3. Wednesday: mean = 38
   $s = \sqrt{(8684/6 - 38^2)} = 1.83$
   Friday: mean = 43
   $s = \sqrt{(11\,110/6 - 43^2)} = 1.63$

4. Mean = 64.4
   $s = \sqrt{(33\,195.54/8 - 64.4^2)} = 1.44\,p$

5. Mean = 8
   $s = \sqrt{(570/8 - 64)} = 2.69$

Self-assessment question 10.3

1. Different weight measurements
   Apples:
       mean = 81
       $s = \sqrt{(99\,245/15 - 81^2)} = 7.44$
       CV = 9.2%
   Oranges:
       mean = 15.4667
       $s = \sqrt{(3690/15 - 15.4667^2)} = 2.60$
       CV = 16.8%
   Orange prices are more variable

Exercise 10.3

1.   mean = 260
       $s = \sqrt{(681\,710/10 - 260^2)} = 23.9$
   CV of A = 9.2%
   CV of B = 7.9%

2. X : 33.3%   Y : 75%

3. mean = 175
   $s = \sqrt{(153\,229/5 - 175^2)} = 4.56$
   CV = 2.6%

Self-assessment question 11.2

1. Strong positive correlation, but note the lack of linearity (see Figure S10)

Figure S10. Absenteeism on workforce

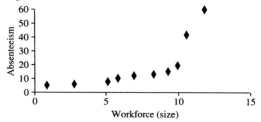

Exercise 11.2

1. Strong negative correlation (Figure S11)

Figure S11. Colour versus black and white TV licences

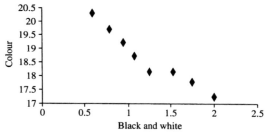

2. Very weak positive correlation (Figure S12)

Figure S12. Assessment marks

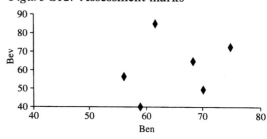

3. Weak negative correlation (Figure S13)

Figure S13. Coursework vs exam

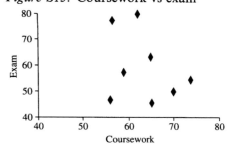

Self-assessment question 11.3

1. $\sum x = 414$, $\sum y = 4960$, $\sum xy = 197\,488$, $\sum x^2 = 18\,260$, $\sum y^2 = 2\,250\,816$
   $r = 316\,416/\sqrt{(47\,724 \times 2\,408\,192)} = 0.93$

Exercise 11.3

1. $\sum x = 56$, $\sum y = 87$, $\sum xy = 572$, $\sum x^2 = 486$, $\sum y^2 = 1281$
   $r = 848/\sqrt{(1724 \times 5241)} = 0.28$

2. $\sum x = 390$, $\sum y = 366$, $\sum xy = 23\,956$, $\sum x^2 = 25\,610$, $\sum y^2 = 23\,572$
   $r = 996/\sqrt{(1560 \times 7476)} = 0.29$

3. $\sum x = 509$, $\sum y = 476$, $\sum xy = 30\,093$, $\sum x^2 = 32\,661$, $\sum y^2 = 29\,556$
   $r = -1540/\sqrt{(2207 \times 9872)} = -0.33$

Self-assessment questions 12.2

1. (a) $\frac{1}{200}$

   (b) $\frac{5}{200} = \frac{1}{40}$

2. (a) $\frac{25}{100} = \frac{1}{4}$

   (b) $\frac{75}{100} = \frac{3}{4}$

3. Probabilities lie between 0 and 1
   Close to zero implies very unlikely to happen

Exercise 12.2

1. (a) $\frac{80}{200} = \frac{2}{5}$

   (b) $\frac{30}{200} = \frac{3}{20}$

   (c) $\frac{120}{200} = \frac{3}{5}$

2. (a) $\frac{8}{40} = \frac{1}{5}$

   (b) $\frac{28}{40} = \frac{7}{10}$

3.

	L	R	Total
G	2	5	7
NG	2	11	13
Total	4	16	20

   (a) $\frac{16}{20} = \frac{4}{5}$

   (b) $\frac{11}{20}$

4. (a) $\frac{3650}{5000} = \frac{73}{100}$

   (b) $\frac{4000}{5000} = \frac{4}{5}$

Index